D0065278

Liquid Dynamics

ACS SYMPOSIUM SERIES **820**

Liquid Dynamics

Experiment, Simulation, and Theory

John T. Fourkas, Editor
Boston College

American Chemical Society, Washington, DC

Library of Congress Cataloging-in-Publication Data

Liquid dynamics : experiment, simulation, and theory / John T. Fourkas, editor.

    p. cm.—(ACS symposium series ; 820)

    Includes bibliographical references and index.

    ISBN 0–8412–3762–X

    1. Solvents—Fluid dynamics—Congresses. 2. Liquids—Fluid dynamics—Congressess.

    I. Fourkas, John T., 1964- II. American Chemical Society. Division of Physical Chemistry. III. American Chemical Society. Meeting (220th: 2000 : Washington, D.C..). IV. Series.

QD544 .L46   2002
541.3′4—dc21   2002016460

Foreword

The ACS Symposium Series was first published in 1974 to provide a mechanism for publishing symposia quickly in book form. The purpose of the series is to publish timely, comprehensive books developed from ACS sponsored symposia based on current scientific research. Occasionally, books are developed from symposia sponsored by other organizations when the topic is of keen interest to the chemistry audience.

Before agreeing to publish a book, the proposed table of contents is reviewed for appropriate and comprehensive coverage and for interest to the audience. Some papers may be excluded to better focus the book; others may be added to provide comprehensiveness. When appropriate, overview or introductory chapters are added. Drafts of chapters are peer-reviewed prior to final acceptance or rejection, and manuscripts are prepared in camera-ready format.

As a rule, only original research papers and original review papers are included in the volumes. Verbatim reproductions of previously published papers are not accepted.

ACS Books Department

Contents

Photochemistry

Vibrational Dynamics

Water

Metastable Liquids

Confined Liquids

Indexes

Preface

Water covers most of the surface of our planet and constitutes a large fraction of the mass of every living organism. Solvents are the favored medium for most chemical synthesis. In these and hundreds of other ways, liquids play a major role in all of our lives. It is ironic, then, that the liquid state of matter is perhaps the most difficult to understand.

This book was inspired by a symposium devoted to the developments in the study of dynamics in liquids that was held at the 220th National Meeting of the American Chemical Society (ACS) in Washington, D.C. on August 20–24, 2000. This symposium brought together researchers from a broad range of disciplines of science and engineering who shared their efforts and discoveries in areas of liquid dynamics that present us with some of the greatest challenges. The first chapter of this book is an introduction to these challenges and some of the new techniques that are being developed to attack them. The remaining chapters recount the cutting-edge research of the participants in these areas. I hope that this book will serve as a good introduction to the problems and the promise in the field for newcomers and as a valuable reference for veterans.

Acknowledgments

I thank the ACS Division of Physical Chemistry, the Petroleum Research Fund, and Coherent Inc. for their support in putting the symposium together, as well as Michael Fayer for his assistance in organizing the symposium. I am also grateful to all of the participants who made the symposium a success, and all the more so to those who went the extra mile to write chapters for this book. I also thank all of the members of the liquids community who refereed these contributions and helped to make the book even better.

John T. Fourkas
Eugene F. Merkert Chemistry Center
Boston College
Chestnut Hill, MA 02467

Introduction

Chapter 1

New Perspectives on Liquid Dynamics

John T. Fourkas

Eugene F. Merkert Chemistry Center, Boston College, 140 Commonwealth Avenue, Chestnut Hill, MA 02467

New techniques in experiment, theory and simulation are leading to rapid progress in our understanding of dynamics in liquids. In this chapter we discuss the goals and challenges in the study of liquid dynamics and we introduce some of the most exciting new developments in this area.

Introduction

From the interiors of living cells to flasks in the laboratory, a tremendous amount of important chemistry occurs in liquid solutions. The solvent can have a major influence on both the rate and the outcome of a solution-phase chemical process. While static properties of the solvent such as polarity and viscosity can affect chemical processes in a significant but relatively predictable manner, it is the ever-changing microscopic environment of liquids that ultimately determines the fate of chemical processes in solution.

Understanding the behavior of liquids on a microscopic level is a challenging problem. Liquids are both highly disordered and dense, which means that most of the approximations that are used to model gases and crystals are not valid in the liquid phase. The structure of a liquid can evolve over a relatively broad range of time and distance scales, which further complicates the problem. Furthermore, the disorder in a liquid means that a solute may lie in any of a range of different local environments that are in constant flux.

Numerous experimental, theoretical, and computational tools have been brought to bear on the problem of liquid dynamics. Major technological advances in all three arenas have opened completely new windows into the study of the dynamics of liquids, and our knowledge is advancing rapidly. In this chapter we will discuss some of the most important outstanding problems in liquid dynamics and review some of what has been learned about these problems recently using a number of cutting-edge techniques.

Outstanding Problems in Liquid Dynamics

Microscopic Dynamics

As mentioned above, the structure of liquids evolves on many time and distance scales. Understanding the connection between the time and distance scales for dynamic processes in liquids is a major challenge. While fast dynamics must necessarily involve only short distance scales, the converse is not necessarily true. As a result, a considerable amount of effort has gone into developing a more detailed picture of the microscopic structure and dynamics of liquids.

One experimental approach to understanding the behavior of liquids at a molecular level is to use spectroscopic techniques to probe fast motions in bulk samples. For instance, techniques such as neutron scattering,[1] low-frequency Raman spectroscopy,[2] and far-infrared spectroscopy[3] are all extremely useful for studying intermolecular vibrations in liquids. Since the frequencies and dynamics of such vibrations are indicative of the distributions of and behavior of the local potentials about individual molecules, these techniques provide a potentially powerful means of exploring liquids on a microscopic distance scale. Furthermore, developments in ultrafast laser technology over the past decade now allow researchers to use time-domain techniques such as optical Kerr effect (OKE) spectroscopy[2] and THz spectroscopy[3] in lieu of frequency-domain Raman and infrared spectroscopies. Coupled with powerful data analysis techniques,[4,5] these time-domain techniques often offer significant advantages over their frequency-domain analogues. A good example of how intermolecular spectroscopy is used in liquids can be found in Chapter 2, in which McMorrow *et al.* demonstrate how OKE spectroscopy can be used to probe the details of intermolecular dynamics in liquids.

A common difficulty with interpreting data from intermolecular vibrations in bulk liquids is that the spectra are broad and generally contain few distinct features. A number of approaches are being taken to improve this situation. On the theoretical and computational side, new tools are being developed for the improved modeling of intermolecular spectra. One such tool is instantaneous normal mode (INM) analysis.[6,7] In this technique, the harmonic modes of a simulated liquid are calculated for many snapshot configurations, and then the properties of these averaged INM densities of states are used to calculate observable quantities. For instance, in Chapter 3 Moore *et al.* use INMs to study the intermolecular spectroscopy of a number of liquids. On the experimental side, new techniques are under development that allow the different contributions to intermolecular spectra to be distinguished from one another.[8,9] While still in their infancy, these techniques hold promise for providing a new window into the nature of intermolecular modes in liquids.

Another common approach to probing microscopic liquid dynamics is to use a guest molecule to report back on its environment. An important example of this type of technique is the time-dependent fluorescence Stokes shift (TRFSS) experiment.[10] In this technique, an ultrafast laser pulse excites a dye molecule into an excited electronic state, and the response of the solvent to this change in the properties of the solute is monitored via the time-dependent behavior of the fluorescence spectrum of the dye molecule following excitation. Beard *et al.* have added an exciting new twist to this technique, and in Chapter 4 they describe their use of THz spectroscopy to monitor the change in the intermolecular infrared spectrum following excitation of a dye molecule.

Guest molecules can also report back on their environment through processes such as diffusion, be it orientational[11,12] or translational.[13] Diffusion studies yield information about local viscosity, the nature of the interactions between the solute and the solvent, local anisotropy, and other microscopic issues. In Chapter 5, Knowles *et al.* discuss an impressive new technique, Fourier imaging correlation spectroscopy (FICS), for studying translation dynamics optically on extremely short distance scales. They demonstrate that this technique is a powerful means of studying dynamics not only in bulk liquids, but also in living cells.

One of the most commonly used techniques in obtaining detailed structural information about solids is X-ray diffraction. Of course, in its normal implementation this technique is not amenable to use in liquids, since the structure of a liquid evolves on a time scale that is much faster than that at which standard X-ray diffraction data can be acquired. This is another realm in which ultrafast laser systems are opening new frontiers, in this case by making possible the generation of X-ray pulses that are fast enough to monitor structure in liquids.[14] Oulianov *et al.* present some new prospects for using X-ray diffraction and spectroscopy to study structure and dynamics in liquids in Chapter 6.

As difficult of a challenge understanding simple liquids might present, a whole new level of complexity exists in the dynamics of polymer melts.[15]

Polymers are such large molecules that intramolecular degrees of freedom can play many of the same roles as do intermolecular degrees of freedom in simple liquids. Many dynamic processes, such as diffusion, necessarily occur by different mechanisms in high-molecular-weight polymers than they do in simple liquids, and many types of apparently anomalous behavior are observed. In her chapter, Guenza uses ideas based on cooperative dynamics to develop a theory that can describe some of these experimental anomalies in Chapter 7.

Photochemistry

In certain molecules, chemical processes such as electron detachment, bond scission, or isomerization can be initiated using light. The dynamics of photochemical processes can change dramatically in going from isolated gas-phase molecules into solution. In some instances, the crowding of the solvent around solutes acts to retard photochemical dynamics, while in others the solvent provides stabilization that can act as a driving force to enhance the rate at which dynamics take place. Indeed, both of these effects can be involved simultaneously.

Developing a microscopic understanding of the events involved in solution-phase photochemical processes is a major challenge in the study of liquids. The study of photochemistry in solution has been facilitated greatly by the development of high-powered, tunable ultrafast lasers, and this book contains a number of excellent examples of what can be achieved with such laser systems.

One important problem in solution-phase chemistry is understanding the behavior of electrons that are ejected from molecules into the surrounding solvent. The ability of an ejected electron to escape from its parent molecule is one important factor in determining the stability of radicals in solution. In many cases an ejected electron effectively bounces off the surrounding solvent cage and recombines with the parent molecule, but under appropriate circumstances the electron can instead make its way into the solvent, where it can be stabilized for a considerable period of time before undergoing a recombination event. Two of the contributions in this volume are concerned with this problem. In Chapter 8 Kloepfer *et al.* describe experiments in which they employ ultrafast lasers to implement ultraviolet photoejection, which is followed by broadband spectral probing to study the photoejection of electrons from aqueous inorganic anions. In Chapter 9 Peon *et al.* explore the dynamics of electrons photoejected from indole and tryptophan into water using ultrafast pump/probe spectroscopy.

Photolysis reactions in solution share many of the attributes of photoionization reactions, such as the competition between geminate recombination and diffusion of the products. The solvent can also play a major role in the distribution of energy in the products of the photodissociation. In

Chapter 10 Philpott *et al.* use time-resolved resonance Raman spectroscopy to explore the dynamics of the photolysis of OClO in ethanol and trifluoroethanol in their chapter. This molecule is an important player in the chemistry of ozone depletion, and ultrafast spectroscopy is proving to be an especially useful tool for unraveling the photochemistry of this species.

Polyenes are an important class of molecules that can undergo photoisomerization. Molecules of this type are responsible for the ability of our eyes to detect light, for instance. It is therefore of great interest to understand how the isomerization of such molecules is affected by solvent. Anderson and Sension employ ultrafast transient absorption spectroscopy to study the excited-state lifetimes and isomerization dynamics of a number of polyenes in Chapter 11. Polyene isomerization is surprisingly unaffected by the solvent in these experiments. This finding reveals important clues as to the nature of the electronic states involved in the isomerization process.

Vibrational Dynamics

Molecular vibrations can also be affected strongly by solvation. The influence of a solvent on the nature of a vibrational mode comes about through a number of mechanisms. For instance, the equilibrium geometry of a molecule may change in solution due to solvent stabilization, which in turn can change the frequencies of its vibrational modes. The distribution of local solvent environments can further lead to a broadening of the vibrational lines. Direct interactions with the solvent can also shift vibrational lines in solution; for example, vibrations that involve large changes in dipole moment may be affected significantly in a polar solvent. The solvent can additionally assist in the redistribution of vibrational energy, thus affecting vibrational lifetimes and line widths.

Recent technological developments in ultrafast lasers are also leaving their mark on infrared and Raman spectroscopies in solution,[16] several notable examples of which are contained in this volume. In Chapter 14 Woutersen and Bakker use ultrafast infrared spectroscopy to study energy transfer among the O-H vibrations of HDO dissolved in D_2O. Their results suggest that mechanisms beyond simple dipole-dipole coupling are responsible for the energy transfer. In Chapter 12 Tominaga *et al.* employ ultrafast Raman spectroscopy to study the behavior of vibrations in liquids confined in small pores. Golonzka *et al.* used frequency-resolved ultrafast infrared pump/probe spectroscopy to determine the structure of rhodium dicarbonyl acetylacetonate in solution in Chapter 13. This last study is an example of how the current generation of laser technology has made it possible to perform multidimensional vibrational (not to mention electronic) spectroscopies in liquids.[17] These developing methods, which are in many ways analogous to multidimensional nuclear magnetic resonance (NMR) techniques, provide a considerable amount of information that cannot be obtained from more traditional spectroscopies. As such, these multidimensional optical spectroscopies are likely to take a central position in the study of liquids in the near future.

Water

Water is, without a doubt, the most important solvent on our planet, and as such its properties are of great scientific interest.[18] It is also one of the most intriguing, and at times mysterious, liquids known. Factors such as the high degree of hydrogen bonding in this liquid lead to a considerable amount of anomalous behavior, including the famous density maximum at 4°C. The dynamics of water are rich in their complexity, and are certain to continue to attract scientific attention for a long time to come.

Many of the contributions in this volume involve water to some extent. Some, such as the above-mentioned Chapter 14 on vibrational energy transfer by Woutersen and Bakker, deal exclusively with this liquid. Smith *et al.* have developed a method for fabricating and then studying films of amorphous glassy water, and use this technique to investigate the connections between glassy water and supercooled water in Chapter 15. This work may have significant implications for our understanding of the phase diagram of water at low temperatures. In Chapter 21, Boyd *et al.* use THz spectroscopy to investigate the properties of water trapped in the interior of inverse micelles. They find that the vibrational density of states of water is altered considerably upon confinement, which has bearing on our understanding of the involvement of water in biological processes.

Metastable States

One of the most fascinating aspects of liquids is their ability to remain in metastable states for considerable periods of time.[19] A significant number of liquids remain stable enough to be studied when superheated, supercooled or stretched. Indeed, metastable liquids can even be stable enough to be used in technological applications. Bubble chambers take advantage of superheated liquids, for instance, and supercooled liquids have found numerous applications.[20]

Because of the large number of materials that can be supercooled, this is undoubtedly the best-studied of the metastable states of liquids.[21,22] In essentially every class of liquids (e.g., hydrogen-bonding, nonpolar, aromatic, etc.), substances can be found that can be supercooled for long periods of time. Despite the vast range of chemical properties found in these substances, many aspects of their behavior are essentially universal. For instance, relaxation in supercooled liquids is increasingly non-exponential as the temperature is decreased, and the temperature dependence of dynamic processes is generally notably non-Arrhenius in nature.

The origin of the non-exponential relaxation in supercooled liquids has been the subject of considerable attention over the years. At one extreme, the relaxation might be homogeneous, which is to say that if one were to observe any microscopic region of a supercooled liquid, the behavior would be identical to that seen in the bulk. At the other extreme, the relaxation could be heterogeneous, in which case each local region could exhibit homogeneous relaxation, but this relaxation would occur on different time scales in different regions. While the nature of supercooled liquids undoubtedly lies somewhere between these extremes, there is mounting evidence that relaxation in supercooled liquids is closer to the heterogeneous extreme than the homogeneous one. Even so, many issues remain to be resolved, such as the size and nature of the local regions in which relaxation is homogeneous and the rate at which the relaxation time scale of a region evolves. Two chapters in this volume are devoted to these problems. In Chapter 17 Chamberlin develops a model of heterogeneity based on the thermodynamics of small systems, and compares his predictions with experimental data on heterogeneity in supercooled liquids. Glotzer *et al.* describe state-of-the art computer simulations of supercooled liquids in Chapter 16. These simulations make it possible to focus on particular dynamic subsets of molecules, such as those undergoing relatively fast motion. Glotzer *et al.* also propose the use of a generalized susceptibility as a indicator of heterogeneous dynamics.

Three other chapters in this volume are concerned with different aspects of supercooled liquids. As discussed above, Smith *et al.* explore the behavior of glassy and supercooled water in Chapter 15. Mohanty employs theoretical methods that were originally developed to treat spin glasses as a means of assessing the inherent nonlinearity of glass-forming liquids in Chapter 18. Sillescu *et al.* use NMR techniques to study intramolecular dynamics in glass-forming liquids in Chapter 19. These authors demonstrate how intramolecular motions can play an important role in dynamics observed near the glass transition.

Confined Liquids

Confined liquids are found in many biological systems and are involved in a vast array of technologies, including lubrication, separations and oil recovery.

When a liquid is confined on distance scales of less that a dozen or so molecular diameters, its behavior can change radically.[23-26] Such changes result from a number of different effects. For instance, proximity to surfaces can lead to enhanced ordering, and thus inhibition of dynamics, in confined liquids. Specific interactions between liquids and surfaces can retard dynamics even further.

There is a considerable effort underway currently to understand the behavior of confined liquids on a microscopic level. Important new tools have been introduced in recent years and applied to this problem. One such instrument is the surface force apparatus (SFA),[27] in which a liquid is placed between two parallel, atomically-flat plates, the distance between which can be controlled with high precision. Recent improvements in SFA technology have further increased its sensitivity, and promise to provide even more insight into the behavior of confined liquids.[28]

Many other techniques have been employed both to confined liquids and to study them. Numerous studies, such as that of Tominaga *et al.* in this volume, have been performed in nanoporous glasses. These materials can be synthesized with relative ease, and feature relatively monodisperse distributions of pore sizes with an average pore size that can be a small as 10 Å. The inverse micelles used by Boyd *et al.* in Chapter 21 are also becoming an increasingly popular system for studying confined water. Inverse micelles are also very monodisperse, and their average size can be controlled readily via the relative mole fraction of surfactant to water.

Huwe and Kremer explore the effects of even more severe confinement in Chapter 20. They place liquids in zeolites of various pore sizes, the smallest of which can only hold a single molecule of liquid per pore. The dynamics of these confined liquids are then probed using broadband dielectric spectroscopy, allowing the authors to determine, from the standpoint of the dynamics to which this technique is sensitive, how many molecules must gather together in order to see liquid-like behavior.

Conclusions

In recent years, tremendous advances have been made on many fronts in the study of liquid dynamics. I have attempted to cover many of the high points in this chapter, but there is a large body of important work that cannot be treated in the space available. I hope that this and the following chapters will give the reader a sense of where we our in our understanding of liquid dynamics, and the extent of the possibilities that the new techniques discussed here will offer in the near future.

Acknowledgments

The author is a Research Corporation Cottrell Scholar and a Camille Dreyfus Teacher-Scholar.

References

(1) Trouw, F. R.; Price, D. L. *Annu. Rev. Phys. Chem.* **1999**, *50*, 571-601.
(2) Fourkas, J. T. *Practical Spectroscopy* **2001**, *26*, 473-512.
(3) Kindt, J. T.; Schmuttenmaer, C. A. *J. Phys. Chem.* **1996**, *100*, 10373-10379.
(4) McMorrow, D.; Lotshaw, W. T. *J. Phys. Chem.* **1991**, *95*, 10395-10406.
(5) Beard, M. C.; Schmuttenmaer, C. A. *J. Chem. Phys.* **2001**, *114*, 2903-2909.
(6) Stratt, R. M. *Acc. Chem. Res.* **1995**, *28*, 201-207.
(7) Keyes, T. *J. Phys. Chem. A* **1997**, *101*, 2921-2930.
(8) Tanimura, Y.; Mukamel, S. *J. Chem. Phys.* **1993**, *99*, 9496-9511.
(9) Fourkas, J. T. *Adv. Chem. Phys.* **2001**, *117*, 235-274.
(10) Barbara, P. F.; Kang, T. J.; Jarzeba, W.; Fonseca, T. In *Perspectives in Photosynthesis*; Jortner, J., Pullman, B., Eds.; Kluwer: Deventer, 1990, pp 273-292.
(11) Tao, T. *Biopolymers* **1969**, *8*, 609-632.
(12) Kivelson, D.; Madden, P. A. *Annu. Rev. Phys. Chem.* **1980**, *31*, 523-558.
(13) Terazima, M. *Acc. Chem. Res.* **2000**, *33*, 687-694.
(14) Guo, T.; Spielmann, C.; Walker, B. C.; Barty, C. P. J. *Rev. Sci. Instrum.* **2001**, *72*, 41-47.
(15) Doi, M.; Edwards, S. F. *The Theory of Polymer Dynamics*; Oxford: Oxford, UK, 1988.
(16) Fayer, M. D., Ed. *Ultrafast Infrared and Raman Spectroscopy*; Marcel Dekker: New York, 2001.
(17) Fourkas, J. T. *Annu. Rev. Phys. Chem.* **2002**, *53*.
(18) Franks, F., Ed. *Water: A Comprehensive Treatise*; Plenum Press: New York, 1972; Vol. 1.
(19) Debenedetti, P. G. *Metastable Liquids: Concepts and Principles*; Princeton University Press: Princeton, NJ, 1996.
(20) Fourkas, J. T. *Chem. and Ind.* **1998**, *16*, 644-650.
(21) Fourkas, J. T.; Kivelson, D.; Mohanty, U.; Nelson, K. A., Eds. *Supercooled Liquids: Advances and Novel Applications*; ACS Books: Washington, 1997; Vol. 676.
(22) Ediger, M. D.; Angell, C. A.; Nagel, S. R. *J. Phys. Chem.* **1996**, *100*, 13200-13212.

11

(23) Drake, J. M.; Klafter, J.; Kopelman, R.; Awschalom, D. D., Eds. *Dynamics in Small Confining Systems*; Materials Research Society: Pittsburgh, 1993; Vol. 290.

(24) Drake, J. M.; Klafter, J.; Kopelman, R.; Troian, S. M., Eds. *Dynamics in Small Confining Systems II*; Materials Research Society: Pittsburgh, 1995; Vol. 366.

(25) Drake, J. M.; Klafter, J.; Kopelman, R., Eds. *Dynamics in Small Confining Systems III*; Materials Research Society: Pittsburgh, 1997; Vol. 464.

(26) Drake, J. M.; Grest, G. S.; Klafter, J.; Kopelman, R., Eds. *Dynamics in Small Confining Systems IV*; Materials Research Society: Warrendale, PA, 1999; Vol. 543.

(27) Granick, S. *Phys. Today* **1999**, *52*, 26-31.

(28) Heuberger, M.; Zach, M.; Spencer, N. D. *Science* **2001**, *292*, 905 908.

Microscopic Dynamics

Chapter 2

Analysis of Intermolecular Coordinate Contributions to Third-Order Ultrafast Spectroscopy of Liquids

Dale McMorrow[1], William T. Lotshaw[2], Joseph S. Melinger[1], and Valeria Kleiman[3,4]

[1]Naval Research Laboratory, Code 6820, Washington, DC 20375
[2]General Electric Research and Development Center, P.O. Box 8, Schenectady, NY 12301
[3]SFA, Inc., under contract to the Naval Research Laboratory, Code 6820, Washington, DC 20375
[4]Current address: Department of Chemistry, University of Florida, Gainesville, FL 32611

The apparently multicomponent, subpicosecond inter-molecular dynamics of liquids are addressed in a unified manner in terms of an inhomogeneously-broadened harmonic oscillator model for a single vibrational coordinate. The model predicts naturally the bimodal character and exhibits good quantitative agreement with the subpicosecond inter-molecular dynamics measured for chloroform and carbon disulfide liquids.

Intermolecular degrees of freedom in liquids exhibit unique spectral and dynamical characteristics that distinguish them from their better understood intramolecular counterparts. Intermolecular spectral lineshapes typically are broad, diffuse, and markedly asymmetric, with the corresponding subpico-second nonlinear-optical (NLO) response functions exhibiting complex decay characteristics. Despite significant effort on the part of both theorists and experimentalists, a satisfactory understanding of intermolecular structure and dynamics of liquids remains elusive.

In this paper we describe a model that accounts surprisingly well for both the qualitative and quantitative characteristics of the subpicosecond third-order nonlinear-optical response of molecular liquids. This model, which is based on an inhomogeneously-broadened distribution of harmonic oscillators, predicts naturally the complex character of the subpicosecond dynamics and, for molecular species of the appropriate symmetry, accurately represents the apparently multicomponent dynamics in terms of a single degree of freedom.

Dynamics of Molecular Liquids

Intermolecular vibrational coordinates differ from their intramolecular counterparts in three significant ways: *i*) the vibrational potential is *defined by* the (mostly nearest neighbor) intermolecular potential, in contrast to intramolecular modes for which the intermolecular potential is a small perturbation; *ii*) the vibrational frequencies span the same timescale as the dephasing and relaxation processes for these modes; and *iii*) local fluctuations in the intermolecular potentials responsible for the decay processes also give rise to distributions in the oscillator frequencies (inhomogeneous broadening) that are of the same magnitude as the oscillator frequencies themselves. These considerations can be expressed by the relation (*1,2*).

$$\omega \approx \omega_c \approx \Delta\omega, \tag{1}$$

where ω is the oscillator frequency, ω_c is a critical frequency related to the 1/e time for fluctuations of the liquid local structure (which give rise to dephasing and relaxation of intermolecular vibrational coordinates), and $\Delta\omega$ is the width of the inhomogeneous frequency distribution of the oscillator.

The qualitative manifestations of these effects are illustrated schematically in fig 1. The spectral density function for an *intra*molecular vibrational resonance is shown in Figure 1a. For the typical intramolecular mode the oscillator frequency (ω_o) is significantly larger than the spectral bandwidth (2σ), giving rise to an isolated resonance. *Inter*molecular degrees of freedom, in contrast, typically occur at frequencies of less than 100 cm^{-1} and, as noted with regard to eq. (1), the dephasing rate typically is of the same magnitude as the oscillator frequency. One significant consequence of this condition, which gives rise to a significant deviation from the symmetrical lineshape that is commonly observed for the intramolecular case, is the interference of the positive and negative frequency contributions in the spectral response. This is

16

illustrated in Figure 1b for a Gaussian lineshape with $\omega_o = \sigma$, but is a general effect that that does not depend on the specific functional form of the lineshape. In practice it is common to plot only the positive-frequency portion of the spectral response illustrated in Figure 1, inverted to produce a positive-valued function.

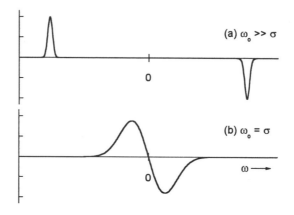

Figure 1. Schematic representation of the Raman spectral density for (a) an isolated intramolecular vibrational mode, and (b) an intermolecular vibrational mode. In this example, ω_o is the center frequency and σ is the standard deviation of the Gaussian function.

The development and maturation of femtosecond nonlinear-optical spectroscopies for probing the structure and dynamics of transparent liquids permits the investigation of intermolecular degrees of freedom to an extent that was not possible previously. There now exists a significant body of work on the third-order ultrafast dynamics of a wide range of molecular liquids (*1-13*). That body of work reveals that each of the molecular liquids investigated to date exhibits an *apparently universal* subpicosecond temporal signature, the characteristics of which are illustrated for three different liquids by the data of Figure 2. Included in Figure 2 are the non-diffusive nuclear contributions to the optical heterodyne detected optical Kerr effect (OHD OKE) impulse response functions of *neat* CS_2, *neat* $CHCl_3$, and a 50% by volume solution of pyridine in CCl_4. Details of the data analysis procedures used to extract these response functions are given below.

Each transient of Figure 2. illustrates: (*i*) an inertially delayed (75 to 150 fs) rise; (*ii*) an initially rapid, Gaussian-like decay; followed by (*iii*) a slower, approximately exponential relaxation. Also apparent in the each of the curves is evidence for a weak recurrence following the initial Gaussian-like decay. In the data of Figure 2 this recurrence is most pronounced for the pyridine solution, but is clearly discernible in the response functions of the other two liquids.

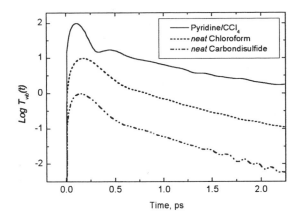

Figure 2. Nondiffusive (vibrational) part of the intermolecular OHD OKE nuclear response function for the three liquids indicated. Measurements were performed at 295 K and the curves are displaced vertically for clarity.

The temporal signature evident in the data of Figure 2 now has been observed in the NLO response of a wide range of *neat* liquids and solutions, and appears to reflect a universal property of low-frequency intermolecular motions in liquids (*1,9,10,13-15*). It is our belief that an understanding of the origin of this characteristic temporal signature will dramatically improve our understanding of the nature of non-diffusive intermolecular degrees of freedom in liquids.

OHD OKE Response of $CHCl_3$

The reduced data representations of Figure 2 are generated using a series of Fourier-transform manipulations that are now well established (*1,9,10,13-5*). In this section we reproduce the key expressions necessary for generating and interpreting the impulse response functions for non-diffusive intermolecular degrees of freedom that are illustrated in Figure 2. The femtosecond OHD OKE response of *neat* chloroform liquid is used to illustrate this procedure.

The optical Kerr effect experiment is a form of pump-probe polarization spectroscopy that interrogates the anisotropic part of the third-order nonlinear-optical susceptibility of the medium. The frequency-domain implementation of the OHD OKE experiment was initially developed in the late 1970s by Levenson and co-workers (*16*), with the time-domain variant of that experiment first implemented in the late 1980s (*3,11*). The data of this study were generated in a conventional OKE experimental geometry (*15*) using the output of a modelocked Ti:sapphire laser that generates 20 − 40 fs nearly-transform-limited optical pulses near 800 nm. With a λ/4 waveplate positioned between the crossed polarizers, a 90° phase-shifted local oscillator is introduced along the path of the probe beam by a slight rotation (< 1°) of the input polarizer. The heterodyne signal is contaminated with the homodyne signal (which scales quadratically with the pump intensity), with the degree of contamination depending on the amplitude of the local oscillator. The pure heterodyne responses are recovered by constructing the sum of data scans collected with positively and negatively sensed local oscillators (*15*).

Within the Born-Oppenheimer approximation (*17*), if all optical frequencies lie below any electronic resonance, the third-order nonlinear-optical impulse response function may be written as a sum of electronic and nuclear parts,

$$R^{(3)}_{\text{OKE}}(t) = \gamma(t) + R_{nuc}(t), \tag{2a}$$

where, for the optical Kerr effect

$$R^{(3)}_{\text{OKE}}(t) = R^{(3)}_{xxyy}(t) + R^{(3)}_{xyxy}(t), \tag{2b}$$

and

$$\gamma(t) = \gamma\, \delta(t) \tag{2c}$$

is the purely electronic hyperpolarizability which is instantaneous on the time scale of the applied laser pulse. For Fourier-transform-limited optical pulses and a local oscillator that is 90° out-of-phase with the probe field, the transmission function of the Kerr cell is given by the convolution integral (*3,15*):

$$T_{\text{OKE}}(\tau) = \int_{-\infty}^{\infty} G_0^{(2)}(\tau - t)\, R_{\text{OKE}}(\tau)\, dt, \tag{3}$$

where $G_0^{(2)}(\tau)$ is the background-free laser pulse intensity autocorrelation function and τ is the delay between pump and probe pulses.

The experimentally measured OHD OKE transmission function, $T(\tau)$, for *neat* chloroform liquid is shown in Figure 3a. Prominent in this data trace is the large amplitude, pulse-limited response centered at $\tau = 0$ that arises from the purely electronic hyperpolarizability, $\gamma(t)$ of eq. (1a). This response is symmetric about $\tau = 0$ and, from eqs. (2) and (3), is represented by a scalar multiple of the intensity autocorrelation function:

$$T_{el}^{(3)}(\tau) = \gamma\, G_0^{(2)}(\tau). \tag{4}$$

All other contributions to the transient of Figure 3a have their origins in the motions of the nuclei, and are contained in $R_{nuc}(t)$. The high-frequency oscillatory contributions arise from the intramolecular vibrational coordinates whose frequencies lie within the coherence bandwidth of the excitation and probing pulses. For chloroform these are the depolarized 262 cm^{-1} ν_6 and the partially polarized 366 cm^{-1} ν_3 deformation modes (18). Also evident, buried under the high-frequency oscillations, is the rise and decay of the intermolecular nuclear response. This response contains the picosecond-timescale relaxation of the laser-induced orientational anisotropy, and the subpicosecond dynamics that arise from dynamically-coherent intermolecular vibrational degrees of freedom.

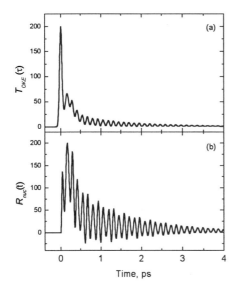

Figure 3. (a) OHD OKE response for neat chloroform liquid measured at 295 K with an out-of-phase local oscillator and (b) its corresponding nuclear impulse response function.

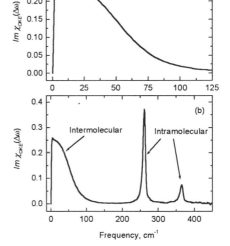

Figure 4. OKE spectral density for neat chloroform liquid corresponding to the data of Figure 3.

The very simple functional form of eq. (3), which is a direct consequence of optical heterodyne detection, lends itself to a range of manipulations that can be used to simplify the presentation and analysis of the experimental data (*14,15*). The Fourier transformation and rearrangement of (3) gives rise to an expression for the (complex) frequency-dependent susceptibility,

$$\chi_{OKE}^{(3)}(\Delta\omega) = \mathfrak{F}\{R_{OKE}(\tau)\} = \frac{\mathfrak{F}\{T_{OKE}(\tau)\}}{\mathfrak{F}\{G_0^{(2)}(\tau)\}},\tag{5}$$

where the first equality of (5) expresses the definition of the frequency-dependent nonlinear-optical susceptibility (*19*), $\Delta\omega = \omega_m \pm \omega_n$, and the $\omega_{m,n}$ are Fourier components of the pump pulse. Since both $G_0^{(2)}(\tau)$ and $T(\tau)$ are experimentally measured functions, $\chi_{OKE}^{(3)}(\Delta\omega)$ is determined from the data with no model-based assumptions. The frequency range over which eq. (5) has sufficient S/N to render meaningful results is determined by the coherence bandwidth of the laser pulses and the quality of the data.

The impulse response function, $R_{OKE}(\tau)$, is a real, causal function of time. Since the intensity autocorrelation function, $G_0^{(2)}(\tau)$ is, by definition, an even function of time, the electronic contribution to the signal (eq. 3) is symmetric and contributes only to the real part of the Fourier transform (5). This result, in conjunction with the causality of the response function, permits the direct determination of nuclear impulse response function from experimentally measured quantities (*14*):

$$R_{nuc}(t) = 2\,\mathfrak{F}^{-1}\{Im\,\chi_{OKE}^{(3)}(\Delta\omega)\}\,H(t\text{-}t_0),\tag{6}$$

where $H(t)$ is the Heaviside step function. All of the information on the nuclear response of the material to the optical pulse is contained in $Im\,\chi_{OKE}^{(3)}(\Delta\omega)$ (*14*), which is shown in Figure 4 for the chloroform data of Figure 3a. The data of Figure 4b illustrate the intermolecular and intramolecular contributions to the spectral density. The nuclear impulse response function, $R_{nuc}(t)$, generated from the spectral data of Figure 4b using eq. (6), is shown in Figure 3b. This data trace illustrates clearly the beating of the two intramolecular modes, which is masked in the raw data (*cf.*, Figure 3a) by the spectral filter effects of the finite bandwidth excitation and probing pulses (*14*).

The primary interest of this study lies with the *intermolecular* contributions to the NLO response function. The frequency response associated with the intermolecular coordinates is illustrated on an expanded scale in

Figure 4a. The clear separation of the intermolecular and intramolecular contributions to spectral density of Figure 4 permits us to take advantage of numerical Fourier filtering techniques to generate independently the intermolecular and intramolecular response functions for chloroform liquid. In the present example we apply a sharp-cut low-pass filter at 175 cm^{-1} to the data of Figure 4 to generate the spectral density function we can call $Im\, \chi_{175}^{(3)}(\Delta\omega)$. The resulting spectral response is identical to that of Figure 4 for frequencies less than ± 175 cm^{-1}, and equal to zero for frequencies greater than ± 175 cm^{-1}.

The application of eq. (6) to the filtered spectral density, $Im\, \chi_{175}^{(3)}(\Delta\omega)$, generates the impulse response function for intermolecular nuclear degrees of freedom in CHCl$_3$, which we refer to as $R_{inter}(\tau)$. Substitution of this response function into eq. (3) generates directly the intermolecular nuclear contribution to the experimental observable:

$$T_{inter}(\tau) = \int_{-\infty}^{\infty} G_0^{(2)}(\tau - t)\, R_{inter}(\tau)\, dt. \tag{7}$$

The intermolecular nuclear OHD OKE response for *neat* chloroform liquid generated using this procedure is given in Figure 5.

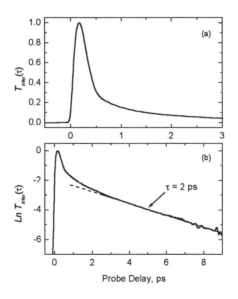

Figure 5. Intermolecular contribution to the OHD OKE response function generated from the chloroform data of Figure 3 using eq. (7).

The response function of Figure 5 includes contributions from dynamically-coherent intermolecular vibrational coordinates and the diffusive relaxation of the laser-induced orientational anisotropy. If we assume separability of the diffusive and vibrational (non-diffusive) contributions to the signal, the nuclear impulse response function can be represented as a sum of two terms:

$$R_{inter}^{(3)}(t) = r_v^{(3)}(t) + r_d^{(3)}(t). \qquad (8)$$

Distinguishing between the diffusive and vibrational contributions as in eq. (8) has been widely utilized in the analysis of femtosecond NLO data, and is usually premised on time-scale separation arguments (3,20). In what follows we assume the validity of (8) and focus on the intermolecular vibrational nuclear contributions, $r_v^{(3)}(t)$, which typically exhibit subpicosecond decay characteristics. This signal component can contain contributions from diverse origins such as molecular librational or free rotational motions, translational modes that lead to a net change in the polarizability (through interaction-induced effects), as well as more complex lattice-like "collective" modes that can involve the reorientation and translation of locally ordered groups of molecules. The generation of $r_v^{(3)}(t)$ requires subtraction of the diffusive contribution from the intermolecular response function. This is achieved using an assumed functional form for $r_d^{(3)}(t)$ and a tail-matching procedure, and is illustrated elsewhere (13). The intermolecular vibrational impulse response functions for $CHCl_3$ and two other liquids generated using this procedure are given in Figure 2.

Inhomogeneous Broadening

An important issue when addressing the nature of intermolecular degrees of freedom is the subject of inhomogeneous broadening. On purely intuitive grounds, given the disordered nature of the liquid state, a significant degree of inhomogeneity in intermolecular coordinates is expected. Basic line-shape analyses of intermolecular Raman spectra and femtosecond NLO transients tend to re-enforce such conclusions (3,7,21,22). It is widely recognized, however, that the information content of coherent Raman spectroscopies, such as the optical Kerr effect, is formally equivalent to that of spontaneous Raman measurements (23-25). As such, it is not possible to discriminate unambiguously between homogeneous and inhomogeneous line-broadening contributions (23-25).

For $\chi^{(3)}$ spectroscopies, then, any conclusions drawn as to the degree of inhomogeneity must be deduced from model-based considerations. Lineshape analyses that are based on specific models for molecular motion can provide insight into the physical processes relevant to the experimental observables. Such analyses have been utilized for many years, with varying degrees of success, to investigate infrared and Raman vibrational spectra, and more recently for the analysis of stimulated Raman spectroscopies, including the optical Kerr effect. The multi-mode curve-fitting analyses noted above, which have provided significant insight into the ultrafast dynamics of molecular liquids, fall into this category.

For an inhomogeneously broadened vibrational coordinate the response function can be represented by a superposition of the different frequency components of the ensemble,

$$R_v^{(3)}(t) = \int\limits_0^\infty d\omega \, g(\omega) \, r^{(3)}(\omega, \, t), \qquad (9)$$

where $r^{(3)}(\omega, \, t)$ represents the homogeneous impulse response function for oscillators at the frequency ω, and the amplitudes of the different frequency contributions are collected in the distribution function, $g(\omega)$. Assuming that the inhomogeneity is a consequence of a Gaussian random process, a Gaussian form for $g(\omega)$ might be expected. In what follows we utilize the function (22),

$$g(\omega) = \frac{1}{2\sigma} \left\{ exp\left[-\frac{(\omega-\omega_o)^2}{2\sigma^2} \right] - exp\left[-\frac{(\omega+\omega_o)^2}{2\sigma^2} \right] \right\}, \qquad (10)$$

where ω_o and σ are parameters which, for an isolated Gaussian distribution, correspond to the center frequency and standard deviation, respectively. Since the function of eq. (10) is not symmetric about ω_o, we refer to ω_o and σ as the *characteristic* frequency and width parameters, respectively. The shape of this function for the case of $\omega_o = \sigma$ is illustrated in Figure 1.

Harmonic Oscillator Model

A representation of the nonlinear-optical response of transparent liquids to electronically nonresonant radiation based on a quantum mechanical harmonic oscillator in the Born-Oppenheimer limit has been developed by Steffen, Fourkas, and Duppen (26). Response functions for third- and fifth-order NLO processes were characterized in the limits of homogeneous and inhomogeneous broadening of the nuclear transitions. Here we apply the results of that work to the analysis of experimental data.

A general expression for the vibrational part of the third-order NLO response function under nonresonant excitation is given by (26),

$$r_v^{(3)}(\tau_1) = \tfrac{1}{\hbar} \sum_{\lambda,\mu} P(\lambda)\, \alpha_{\lambda\mu}(q)\, \alpha_{\mu\lambda}(q)\, sin(\omega_{\lambda\mu}\tau_1), \tag{11}$$

where $P(\lambda)$ denotes the equilibrium distribution of eigenstates $|\lambda\rangle$, the dependence of the polarizability on the nuclear coordinates "q" is explicitly indicated, τ_1 is the delay between the pump and probe laser pulses and, for simplicity, the tensorial notation has been suppressed. This result does not depend on the specific forms of the Hamiltonian or the polarizability operator (26).

Calculation of the matrix elements $\alpha_{\lambda\mu}$ requires the adoption of a model for nuclear motion. For harmonic motion the nonlinear-optical response function can be transformed into terms of the HO eigenstates $|\lambda\rangle$. Under the constraint of damping in the weak coupling limit, and considering only the linear dependence of the polarizability on nuclear coordinates, eq. (11) becomes (26)

$$r_v^{(3)}(\tau_1) = -\frac{\alpha_1^2}{2m\omega} sin(\omega\tau_1) \sum_{\lambda} P(\lambda) \left[\lambda e^{-\Gamma_{\lambda,\lambda-1}\tau_1} - (\lambda+1)e^{-\Gamma_{\lambda,\lambda+1}\tau_1} \right], \tag{12}$$

where α_1 is defined by the polynomial expansion $\alpha(q) = \alpha_1 q + \alpha_2 q^2 + \cdots$, When damping is assumed to be independent of quantum number (level independent damping), eq. (12) reduces to a particularly simple form (26),

$$r_v^{(3)}(\omega,t) = \frac{\alpha_1^2}{2m\omega}\, e^{-\Gamma t}\, sin(\omega t), \tag{13}$$

for which dephasing is fully described by a single decay rate of the form,

$$\Gamma = \gamma + \Gamma^*. \tag{14}$$

In this expression, γ is the population decay rate, Γ^* is the pure dephasing rate and, for the present analysis we assume that Γ is frequency independent.

Application of the inhomogeneously-broadened harmonic-oscillator model to the $CHCl_3$ intermolecular vibrational response is shown in Figure 6. An analogous analysis for *neat* CS_2 data is presented in Figure 7. The fitting procedure involves three parameters: ω_o and σ of the distribution function (eq. 10), and Γ of the oscillator impulse response function (eq. 13).

As is evident for both liquids, the HO oscillator model for a single nuclear coordinate does an excellent job of describing the temporal dynamics of the

OKE vibrational response function at all probe delays. The inhomogeneous oscillator model predicts naturally the key qualitative characteristics of the liquid dynamics: the initial Gaussian-like decay followed by a slower, quasi-exponential relaxation, and exhibits very good quantitative agreement as well. A particularly satisfying aspect of this analysis is the ability to accurately represent the complex, apparently multicomponent subpicosecond dynamics with a response function for a *single intermolecular vibrational coordinate*. Analytical descriptions of the bi-modal character of the dynamics require the superposition of at least two independent coordinates.

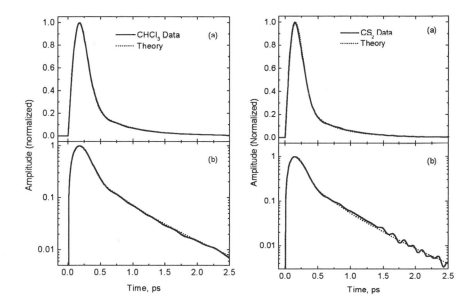

Figure 6. Nondiffusive intermolecular OHD OKE impulse response function for chloroform together with a best fit to the inhomogeneously-broadened HO model.

Figure 7. Nondiffusive intermolecular OHD OKE impulse response function for CS$_2$ together with a best fit to the inhomogeneously-broadened HO model.

The results of figs. 6 and 7 make a strong statement with regard to the question of inhomogeneity in liquids. To the extent that the intermolecular vibrational dynamics of liquids can be described in terms of a linearly independent collection of oscillators (harmonic or otherwise), the presence of the characteristic temporal signature that is observed for CHCl$_3$, CS$_2$, and numerous other liquids, *is a direct indication that the mode in question is inhomogeneously broadened.* Stated conversely, the characteristic temporal signature of intermolecular modes in liquids, a fast Gaussian-like decay followed by a slower quasi-exponential relaxation, can not be accounted for in terms of exponentially damped sinusoids without the consideration of inhomogeneous broadening. This conclusion has been suggested previously in the context of a similar oscillator model based on classical mechanical oscillators (*1,2*). That model, however, fails to provide the quantitative agreement that is evident in figs. 6 and 7 (*27*). In view of the analysis presented above, the alternate possibilities of two independent modes, or a single homogeneously-broadened degree of freedom with the required two-component relaxation dynamics, respectively invoke a complexity in the liquid structure that is inconsistent with the symmetry of either $CHCl_3$ or CS_2, and a response function that cannot be supported by a conventional equation of motion.

The behavior of this model can be better understood by investigating the characteristics of the impulse response function, which is given by eq. (13). This function is plotted in Figure 8 as a function of ω for fixed Γ. For constant Γ, as the oscillator shifts to lower frequency ($\omega < \Gamma$), the impulse response function approaches what would be called, in a classical oscillator, the critically damped limit. In this case eq. (13) can be written as

$$\lim_{\omega \to 0} r_n^{(3)}(\omega,t) = r_n^{(3)}(t) = \frac{\alpha_1^2}{2m}\, te^{-\Gamma t}. \tag{15}$$

As is evident, in this limit, the impulse response function is independent of frequency and depends exclusively on the dephasing rate, Γ. Thus, to the extent that Γ is constant, all oscillators for which the limiting condition of $\omega \ll \Gamma$ is fulfilled will *exhibit identical OKE response functions.* As is illustrated in Figure 8, this limiting case is attained for $\omega \approx \Gamma/4$, but the oscillators begin to take on critically-damped character as ω becomes less than Γ.

Referring to the data of Figure 8, then, the origin of the characteristic temporal signature of molecular liquids becomes clear. For inhomogeneously-broadened coordinates, the higher-frequency more weakly damped components ($\omega \ll \Gamma$) undergo destructive interference, resulting in the rapid, Gaussian-like decay. This is strictly analogous to the inhomogeneous broadening of intramolecular modes. For intermolecular modes, however, the oscillator distribution typically extends to low enough frequencies that the condition $\omega < \Gamma$ is simultaneously fulfilled. These "oscillators" do not actually oscillate, but rather interfere constructively at all times, giving rise to the quasi-exponential tail. This latter effect is unique to intermolecular vibrational coordinates (*cf.*, eq. 1), and is responsible for the bimodal character of the intermolecular spectra and dynamics.

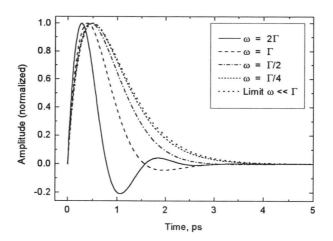

Figure 8. Behavior of equation (13) plotted as a function of ω for fixed Γ. Note the limiting case for small ω.

Conclusions

An inhomogeneously-broadened harmonic oscillator model of inter-molecular nuclear coordinates is applied to the analysis of subpicosecond optical Kerr effect dynamics of *neat* $CHCl_3$ and CS_2 liquids. For an inhomogeneously-broadened distribution of oscillators, this model predicts naturally the bimodal character of the subpicosecond OKE dynamics, a Gaussian-like ultrafast decay and a slower, quasi-exponential relaxation, that is universally observed in molecular liquids. The description of the intermolecular dynamics of the highly-symmetric $CHCl_3$ and CS_2 molecules in terms of *a single vibrational coordinate* is quite appealing, eliminating the need to assign the intermediate-lifetime, quasi-exponential relaxation to a unique nuclear coordinate (as is necessary in multi-mode curve fitting approaches). For highly symmetric molecular species, such as CS_2, such assignments are questionable, at best.

Furthermore, within the harmonic oscillator representation of the nuclear response functions, inhomogeneous broadening of the nuclear coordinate is *required* to achieve this result; the presence in the OKE data of the characteristic Gaussian-like ultrafast decay followed by a slower, quasi-exponential relaxation, is a clear indication that the intermolecular dynamics are inhomogeneously broadened.

In addition to the liquids presented here, we also have investigated the OKE response of binary solutions if CS_2 in alkane solvents. The spectral evolution effects that are observed on dilution are well accounted for by the present model (27). At lower temperatures the OKE response of molecular liquids acquires a more complicated form. A preliminary analysis of the temperature-dependent OKE dynamics of CS_2 liquid provides additional support for the general applicability of the inhomogeneously-broadened oscillator approach (28).

Acknowledgments

The authors wish to acknowledge numerous insightful conversations with John Fourkas. This work is supported by the Office of Naval Research.

References

1. McMorrow, D.; Lotshaw, W.T. *Chem. Phys. Lett.* **1991**, *178*, 69.
2. McMorrow, D.; Thantu, N.; Melinger; J.S.; Kim, S.; Lotshaw, W.T. *J. Phys. Chem.* **1996**, *100*, 10389.
3. McMorrow, D.; Lotshaw, W.T.; Kenney-Wallace, G.A. *IEEE J. Quant. Elect.* **1988**, *QE-24*, 443.
4. Kalpouzos, C.; McMorrow, D.; Lotshaw, W.T.; Kenney-Wallace, G.A. *Chem. Phys. Lett.* **1988**, *150*, 138; *ibid.* **1989**, *155*, 240.
5. Chang, Y.J.; Castner, E.W. Jr. *J. Chem. Phys.* **1993**, *99*, 113.
6. Chang, Y.J.; Castner, E.W. Jr. *J. Chem. Phys.* **1993**, *99*, 7289.
7. Chang, Y.J.; Castner, E.W. Jr. *J. Phys. Chem.* **1996**, *100*, 3330.
8. Neelakandan, M.; Pant, D.; Quitevis, E.L. *J. Phys. Chem.* **1997**, *101*, 2936.
9. Loughnane, B.J.; Scodinu, A.; Farrer, R.A.; Fourkas, J.T. *J. Chem. Phys.* **1999**, *111*, 2686
10. Steffen, T.; Meinders, N.A.C.; Duppen, K. *J. Phys. Chem.* **1998**, *102*, 4213.
11. Lotshaw, W.T.; McMorrow, D.; Kalpouzos, C.; Kenney-Wallace, G.A. *Chem. Phys. Lett.* **1987**, *136*, 323.
12. Lotshaw, W.T.; McMorrow, D.; Kenney-Wallace, G.A. *Proc. SPIE* **1988**, *981*, 20.
13. McMorrow, D.; Lotshaw, W.T. *J. Phys. Chem.* **1991**, *95*, 10395.
14. McMorrow, D.; Lotshaw, W. T. *Chem. Phys. Lett.* **1990**, *174*, 85; McMorrow, D. *Opt. Comm.* **1990**, *86*, 236.

15. Lotshaw, W.T.; McMorrow, D.; Thantu, N., Melinger, J.S.; Kitchenham, R. *J. Ram. Spect.* **1995**, *26*, 571.
16. Eesley, G.L.; Levenson, M.D.; Tolles, W.M. *IEEE J. Quant. Elect.* **1978**, 14, 45; Levenson, M.D.; Eesley, G.L. *Appl. Phys.* **1979**, *19*, 1.
17. Hellwarth, R.W. *Prog. Quant. Elect.* **1977**, *5*, 1.
18. Herzberg, G. *Infrared and Raman Spectra of Polyatomic Molecules;* Van Nostrand Reinhold: New York, 1950.
19. Butcher, P.N. *Nonlinear Optical Phenomena;* Ohio State University: Columbus, 1965.
20. Madden P.A. in *Ultrafast Phenomena IV;* Springer: Berlin, 1984; pp. 224-251.
21. Kalpouzos, C.; McMorrow, D.; Lotshaw, W.T.; Kenney-Wallace, G.A. *J. Phys. Chem.* **1987**, *91*, 2028
22. Friedman, J.S.; She, C.Y. *J. Chem. Phys.* **1993**, *99*, 4960.
23. Loring, R.F.; Mukamel, S. *J. Chem. Phys.* **1985**, *83*, 2116.
24. Tanimura, Y.; Mukamel, S. *J. Chem. Phys.* **1993**, *99*, 9496.
25. Mukamel, S. *Principles of Nonlinear Optical Spectroscopy* (Oxford University Press, New York, 1995).
26. Steffen, T.; Fourkas, J.T.; Duppen, K. *J. Chem. Phys.* **1996**, *105*, 7364; Steffen, T.; Duppen, K. J. Chem. Phys. **1996**, *106*, 3854.
27. McMorrow, D.; Thantu, N.; Kleiman, V.; Melinger, J.S.; Lotshaw, W.T. *J. Phys. Chem. A* **2001**, *105*, 7960.
28. McMorrow, D.; Lotshaw, W.T.; Fourkas, J.T.; Loughnane, B.J.; and Farrer, R.A. *manuscript in preparation.*

Chapter 3

A Combined Time Correlation Function and Instantaneous Normal Mode Investigation of Liquid-State Vibrational Spectroscopy

Preston B. Moore[1], Heather Ahlborn[2], and Brian Space[2,*]

[1]Department of Chemistry, University of Pennsylvania, 231 South 34th Street, Philadelphia, PA 19104–6323
[2]Department of Chemistry, University of South Florida, 4202 East Fowler Avenue, CHE 305, Tampa, FL 33620–5250

Abstract

Our work investigates the instantaneous normal mode (INM) and time correlation function (TCF) of several different liquids and compares the results with the experimental infrared (IR) and Raman spectra. Our work demonstrates that INM and TCF methods can be used in a complementary fashion in describing liquid state vibrational spectroscopy. INM derived spectra often lead to intermolecular spectra that are in agreement with corresponding TCF results, suggesting that the inter-molecular dynamics can be interpreted as oscillations. TCF derived spectra, while formally exact in the low frequency regime ($\hbar\omega \ll kT$), suffer severely from the need for quantum (detailed balance) correction at higher frequencies.

Our approach is to compare TCF spectra with experiment to establish that our MD methods can reliably describe the system of interest, and to employ INM methods to analyze the molecular and dynamical basis for the observed spectroscopy. We have been able to elucidate, on a molecularly detailed basis, a number of condensed phase line shapes, most notably the origin of the broad and intense, high frequency intramolecular O-H stretching absorption in liquid water. This approach has proven successful in systems as diverse as liquid CS_2 and H_2O at a number of state points.

1   Background

Instantaneous normal mode (INM) theory has been successful in describing a number of liquid state properties.[1, 2, 3, 4, 5, 6, 7, 8, 9, 10, 11] Further, INM theory is an intuitively appealing and effective approach to modeling condensed phase spectroscopy.[12, 13, 14, 15, 16, 17, 18, 19, 20, 21, 22, 23, 24, 25, 26] The INM approach to spectroscopy can be interpreted as an approximation to the true Born-Oppenheimer vibrational density of states (DOS).[27] Spectroscopic quantities may be evaluated by making an harmonic approximation to the relevant frequency domain golden rule expression, leading to an approximation to the spectrum in the form of a weighted DOS (wDOS). For IR (or Raman) spectroscopy this becomes a squared dipole (or polarizability) derivative weighted INM DOS, where the derivative of the dipole is with respect to the INM's. The INM wDOS can be constructed from thermodynamic information, not requiring explicit dynamical input, although it contains short time dynamical information. Consequently, it has been argued that motional narrowing effects are a major part of what is neglected in this approximation, and the inclusion of motional narrowing effects into INM theory is the subject of ongoing research.

In several dense liquids, where rotations are significantly hindered, INM intermolecular spectra have been shown to be almost equivalent to time correlation function (TCF) derived spectra,[12, 13, 14, 15, 19, 22, 28] and this has led to an interpretation of the intermolecular dynamics in the slow modulation limit of Kubo's motional narrowing theory.[29] Intramolecular line shapes can be interpreted as being in the motionally narrowed, fast modulation limit, where the underlying INM "spectral density" is narrowed into a slimmer line shape. The Kubo theory of motional narrowing focuses on the line shape associated with a time dependent harmonic oscillator, and it becomes simple in the limits of slow and fast modulation. In the slow modulation limit the line shape reflects all the frequency fluctuations that the oscillator exhibits; in the fast modulation (motionally narrowed) regime the line shape is narrowed. In interpreting an INM wDOS in this context one assumes that INM's are well defined for some time[18, 30] and that they fluctuate in frequency. This kind of interpretation is suggested by the time dependence of the INM frequencies.

Figure 1 shows the time dependence of INM frequencies for liquid CO_2 and H_2O , as they evolve in time. A number of properties are apparent from the figure. One is that the INM frequencies intuitively reflect the vibrational structure of the liquid. There are well defined bands of vibrations with varying widths, reflecting the coupling of the vibrational motion to its surrounding. In liquid CO_2 the antisymmetric stretching

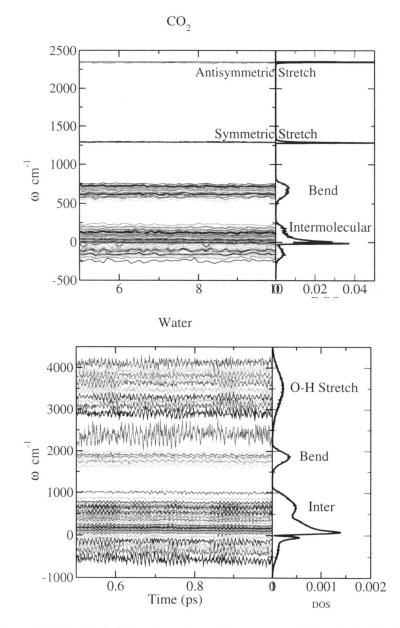

Figure 1: The the left the time dependence some of the adiabatic INM's for liquid CO_2 (top) and H_2O (bottom) is shown. On the right is the DOS of each system, the imaginary INM's are shown on the negative frequency axis.

CS$_2$ Reduced Raman

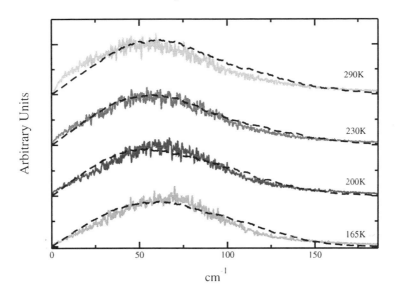

Figure 2: A comparison between the INM (gray noisy) and TCF (black dash) derived reduced Raman spectra for liquid CS_2 at different temperatures. Imaginary INM's are contributing significantly to the INM signal. The theoretical results agree nearly quantitative with the experimental results[31] at all temperatures.[12]

motion is very narrow while the bend, at a much lower frequency, is significantly broadened, and the intermolecular region is the broadest. However in water, while the bend is still a well defined band in the condensed phase, antisymmetric and symmetric stretching bands are no longer distinct. This is consistent with the experimental observation of a single broad O-H spectroscopic line shape. What is not apparent from the figure is the identity of the modes from time slice to time slice.[18, 30] Identifying the modes at each step is necessary to include motional narrowing effects into INM theory of spectroscopy. However, if motional narrowing effects are neglected (the slow modulation limit) it is only necessary to perform an ensemble average over INM's and configurations.[17, 18] Also clear from figure 1 is that there is a persistent band of imaginary INM's. Thus, although it might be supposed that the lifetime of an (instantaneously) unstable motion is short, the amount that exists at any one time is roughly constant. The liquid constantly exhibits unstable motions over time, and we have argued that these motions lead to infrared absorption and Raman scattering.[12, 14, 19]

Consistent with the Kubo picture, it has been found that the intramolecular INM bands are broader but have the nearly the same central frequency as the corresponding TCF line shape, and the integrated intensities of the TCF and INM bands are approximately equal.[13, 15, 18, 20, 22, 29, 32] These two observations suggest the identification of the INM wDOS as an underlying spectral density, *i.e.* the intramolecular line shape with motional narrowing effects removed.

A major strength of the INM method is that it is a frequency domain method, allowing for the modes at any given frequency to be analyzed in molecular detail. INM theory provides a tractable method for examining the underlying molecular motions responsible for generating the associated INM and presumably TCF spectroscopic signature. This is possible because, as noted above, INM intermolecular spectra are often in near agreement with corresponding TCF results; when concerned with intramolecular line shapes it is argued, that the INM spectrum still represents the underlying "spectral density" that is motionally narrowed to give the observed line shape. This agreement, however, does depend on approximately including the imaginary INM's in the spectra, and this issue will be discussed below.[12, 18]

2   Intermolecular Spectroscopy: CO_2 and CS_2

To date, INM methods have provided an excellent description of the infrared intermolecular spectroscopy of CO_2,[18] CS_2 ,[19, 20] and water,[13, 15] along with the Raman spectrum of CS_2 [12, 14] at several state points, and molten $ZnCl_2$.[22] The near agreement between INM and TCF intermolecular spectra suggests that the intermolecular dynamics can be thought of as being described by oscillators that are accurately represented in the slow modulation limit (at all but the lowest frequencies).[14, 18, 22, 29, 32, 33] In this case, all the frequency fluctuations exhibited by the liquid, captured naturally by an INM calculation, manifest themselves in the actual spectrum. This correspondence between the INM and TCF spectra permits the direct examination of the types of molecular motions (*e.g.* rotations,translations, stable, or instantaneously unstable) that are responsible for a spectroscopic signal at a particular frequency. Figure 2 displays a comparison between INM and TCF calculations of the temperature dependent reduced Raman spectrum of CS_2 ; the agreement is remarkable. It should be stressed that the agreement between the curves is absolute in the sense of no parameters are being adjusted to compare the two. This is strong evidence that there is merit to the interpretation of INM spectra in the context of Kubo's theory; the intermolecular spectra are simply reflecting all the frequency fluctuations exhibited as the liquid evolves dynamically.

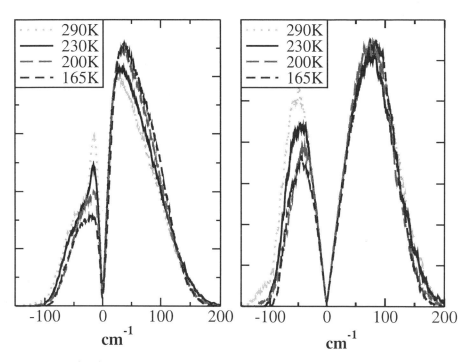

Figure 3: (left) The INM (unweighted) DOS's are shown at the different temperatures considered. The DOS is normalized in all cases. (right) The INM weighted DOS's (wDOS) are shown. The wDOS is presented as an absolute intensity, and would represent the INM reduced Raman spectrum if the imaginary modes were included in the signal *via* our prescription.

At lower frequencies, the agreement between INM and TCF spectra depends critically[12] on contributions from imaginary INM's. In earlier papers we proposed that the imaginary INM's contribute to a spectrum like the real INM's, but at the magnitude of their imaginary frequency, and our idea has been discussed previously.[14, 18] Briefly, the model that was adopted to deduce the contribution of the imaginary INM's was that of a harmonic barrier crossing. We proposed that barrier crossing occurs periodically, *e.g.* a coordinate is being buffeted back and forth over a barrier (this is appropriate in considering the dense liquids that have been examined). Evidence for this type of motion can be seen in Figure 1 where there is a constant imaginary frequency mode. The trajectory of a coordinate crossing a barrier, can then be expanded in a Fourier series, and the TCF evaluated for the coordinate. The leading term in the expansion leads to absorbance at approximately the magnitude of the imaginary frequency.

To further elaborate this point, Figure 3 compares the temperature dependent (unweighted) INM DOS's and the weighted DOS's (wDOS); the INM approximation to the spectroscopy shown in Figure 2 are the wDOS with the imaginary portion allowed to contribute on the real axis at the magnitude of the frequency. The wDOS's are given by Equation 2.1, with the imaginary density plotted on the negative abscissa. When the imaginary INM contribution is included, Equation 2.1 is the INM approximation to the reduced Raman spectrum, $R(\omega)$.

$$R(\omega) = \frac{\hbar}{2} < \delta(\omega - \omega_i) \left[\mathrm{Tr}(\beta' \cdot \beta') \right] > \qquad (2.1)$$

Above, β is the traceless anisotropic portion of the polarizability tensor and Tr represents the trace of the matrix product. In Equation 2.1 β' represents the derivative of the elements of β with respect to the i'th INM, that has a frequency ω_i . The angle brackets represent an ensemble average over liquid configurations. In examining the DOS, it is clear that the imaginary portion grows with temperature. This is expected, especially because at least some of the imaginary INM's describe diffusive modes that would be increasingly important at higher temperatures. Further, the fraction of imaginary INM's in the DOS can be used to formulate a very successful theory of the diffusion coefficient in which the diffusion coefficient is roughly proportional to the fraction of imaginary INM's present, especially when one considers purely translational INM's.[1]

Figure 3 compares the DOS with the wDOS's. It is striking that the real lobe of the wDOS (the comparison is again of absolute intensities in arbitrary units), is virtually temperature independent. The shape of the real wDOS is also different than the DOS itself. The real portion of the DOS decreases with increasing temperature, while the wDOS does not change. However, the imaginary wDOS is changing drastically with temperature, both increasing in magnitude and tending toward higher frequencies. The integrated contributions of the imaginary INM's to the reduced Raman spectrum are 25% at 165K, 26% at 200K, 30% at 230K, and 34% at 290K. Further, the difference in shapes of the curves is what leads to the agreement between the INM and TCF spectra. This is suggestive evidence that motions corresponding to the imaginary frequency INM's are largely responsible for the temperature dependence of the reduced Raman spectrum (or equivalently the OKE spectral density given that they are simply related).[12] This conclusion follows because when these imaginary contributions are included in constructing our INM spectra, the agreement (Figure 2) with temperature dependent TCF results is outstanding, but without imaginary contributions the INM spectra would be essentially temperature independent. This is quite a surprising conclusion given the typical interpretation of low frequency spectroscopy in terms of collective stable oscillations, *e.g.* experiencing a friction due to the other bath coordinates such as in a Brownian oscillator description.[34] While the INM approximation to the spectroscopy certainly misses some dynamical effects, and our inclusion of the imaginary INM's is quite approximate, it is remarkable how closely the INM description mimics its TCF counterpart, and what interesting physical insights it yields.

3 Intramolecular Spectroscopy: Water

The spectroscopy of water has been extensively examined experimentally and theoretically by a variety of methods.[35, 36, 37, 38, 39, 40, 41, 42, 43, 44, 45] We have demonstrated that simple classical molecular dynamics simulations in conjunction with a reasonable model of the system dipole and polarizability are capable of nearly quantitative reproducing most of the intermolecular and intramolecular IR spectroscopy of normal and heavy water.[13, 15] Figure 4 compares INM, TCF, and experimental spectra, all in absolute units. While the agreement in the intermolecular region is very good, it is not especially surprising.[46] However, the agreement in the O-H stretching region is remarkable and unexpected. The resonance is complex and ostensibly highly quantum mechanical and anharmonic. The dynamics are modeled classically to calculate the TCF and the agreement is achieved after quantum correcting with a function appropriate for harmonic dynamics. Further, condensed phase absorptions typically integrate to the value of the gas phase squared dipole derivative. This is the case with the water bending absorption, and this usually indicates that polarization effects are unimportant. In contrast, the water O-H stretching absorption increases by factor of eighteen from the gas phase. This increase in intensity is due to many body polarization effects on the hydrogen bonded O-H stretching coordinate.[15, 36, 37]

Because the TCF results capture both the line shape and intensity of the O-H stretch absorption, certain conclusion can be drawn about the underlying dynamics. The classical TCF spectrum must be corrected to obey detailed balance, and different corrections have been used in the literature.[27] These corrections vary dramatically in their high frequency ($\hbar\omega \gg kT$) behavior, from a constant to exponentially increasing. In the case of the water O-H stretching spectrum, only the harmonic oscillator correction (used in the figure), which makes the absorption for classical and quantum harmonic oscillators the same,[47, 48] gives a physically reasonable result, and even gives near quantitative perfect agreement. This

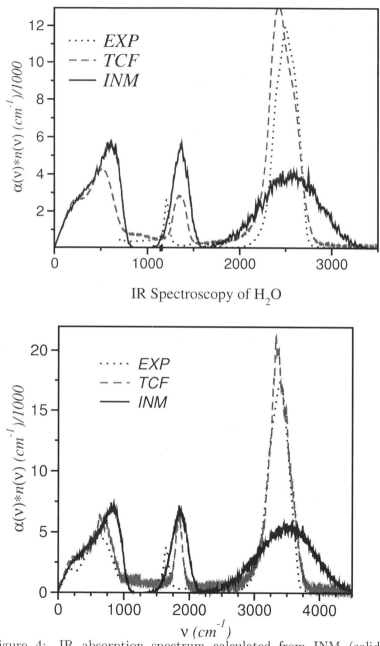

Figure 4: IR absorption spectrum calculated from INM (solid lines), TCF (dash) and reported experimentally (EXP (dotted)) for heavy water D_2O (top) and water H_2O (bottom).

suggests the underlying motions responsible for the absorption are effectively harmonic.

OH Stretch

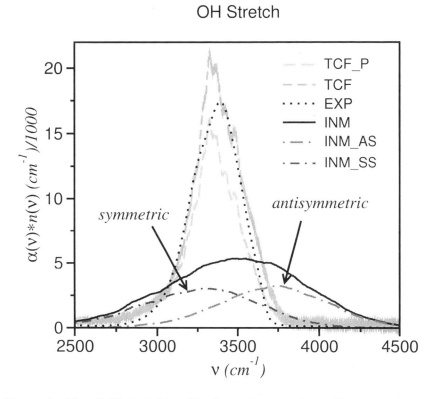

Figure 5: The O-H stretching IR absorption spectrum from experiment, TCF and INM. The black dots are the experimental results. The two dash gray lines are the TCF results, the short dash is the total TCF where the long dash TCF_P is the contribution from the induced polarization effects. The solid gray lines are the INM, which is further broken down into the antisymmetric (INM_AS) and symmetric (INM_SS) contributions.

Figure 5 focuses on the O-H stretching region. The graph shows the total absorption as well as the many-body polarization components. The total TCF is a sum of the permanent charge and polarization components, figure 5 clearly demonstrates that the OH stretch is dominated by the polarization enhancement. This polarization effect is missing in the gas phase. The figure also shows the INM spectrum, that has approximately the same central frequency and integrated intensity, but is broader, consistent with the motional narrowing picture that was described above. The INM spectrum is projected onto the gas phase symmetric and antisymmetric stretches to ascertain their condensed phase contribution. It is clear that the condensed phase motions are a mixing of the gas phase normal modes. In fact, the condensed phase normal motions are nearly simple O-H bond stretching motions.[13] This point of view is further supported by the fact that the INM DOS of states can be essentially reproduced by simply making a histogram of the O-H local mode frequencies (by noting the instantaneous bond lengths).[13]

The O-H stretching potential is quite anharmonic and liquid water molecules are found in a variety of surroundings. Water molecules are thus distorted by the local environment away from their gas phase equilibrium bond lengths. The picture that emerges from these observations is that the O-H bonds then vibrate nearly harmonically around this "distorted" geometry. This simple effectively harmonic motion is a result of a complex gas phase potential being placed into a complicated condensed phase environment, with emergent simplicity. While quantum corrected classical TCF calculations may not be appropriate for modeling all intramolecular motions in liquids, when they work well they provide tremendous insight into the dynamical system.

4 Conclusion

We have demonstrated that INM and TCF can be used in a complementary fashion to gain insight into the dynamics and spectroscopy of molecular liquids. INM and TCF can be used to investigate the IR and Raman spectroscopies from the far infrared to the highly quantum mechanical 3000 cm^{-1} OH stretch in water.

In examining the low frequency intermolecular spectroscopy of liquid CO_2 and CS_2, strong evidence was also obtained that, at least instantaneously, unstable motions make a significant ($\approx 25\%$) contribution to the integrated intermolecular spectroscopy for a number of different state points. The results have also demonstrated that, using the power of modern computation, careful TCF calculations can nearly quantitatively reproduce experimentally measured infrared and Raman or optical Kerr effect (OKE) spectra using relatively simple molecular dynamics (MD) force fields and spectroscopic (system dipole and polarizability) models. This approach has proven successful in systems as diverse as liquid CS_2 and H_2O at a number of state points.

We were also able to demonstrate that combined INM and TCF investigation of the O-H stretching region in water is possible with the present methods. The next goal is to be able to model the structure and dynamics of water in more complex environments such as air - water and water - carbon tetrachloride interfaces.[49, 50] We will again take the approach of validating our models by detailed comparison with (sum frequency generation) spectroscopy, and investigation of the systems *via* our combined approach. INM spectroscopic methods would also be useful in modeling biopolymers, where it is crucial to associate a spectroscopic signature with the conformation of the molecule.

5   Acknowledgments

Brian Space and Preston Moore would like to thank Professor M.L. Klein for continuing encouragement in their collaboration. The research at the University of South Florida was supported by an NSF Career Grant (No. CHE-9732945) to Brian Space.

References

[1] W.-X. Li and T. Keyes, J. Chem. Phys. **111**, 5503 (1999).

[2] J. Jang and R. M. Stratt, J. Chem. Phys. **112**, 7524 (2000).

[3] R. E. Larsen and R. M. Stratt, J. Chem. Phys. **110**, 1036 (1999).

[4] R. L. Murry, J. T. Fourkas, W.-X. Li, and T. Keyes, Phys. Rev. Lett. **38**, 3550 (1999).

[5] T. Keyes, J. Phys. Chem. **101**, 2921 (1997).

[6] R. M. Stratt, Acc. Chem. Res. **28**, 201 (1995).

[7] J. Adams and R. Stratt, J. Chem. Phys. **93**, 1632 (1990).

[8] B. Ladanyi and R. Stratt, J. Chem. Phys. **99**, 2502 (1995).

[9] T. Keyes, J. Chem. Phys **104**, 9349 (1996).

[10] P.B. Moore and T. Keyes, J. Chem. Phys **100**, 6709 (1994).

[11] P.B. Moore, A. Tokmakoff, T. Keyes, and M. Fayer, J. Chem. Phys **103**, 3325 (1995).

[12] X. Ji, H. Ahlborn, B. Space, and P.B. Moore, J. Chem. Phys. **113**, 8693 (2000)

[13] H. Ahlborn, X. Ji, B. Space, and P. B. Moore, J. Chem. Phys. **112**, 8083 (2000).

[14] X. Ji, H. Ahlborn, B. Space, P.B. Moore, Y. Zhou, S. Constantine, and L. D. Ziegler, J. Chem. Phys. **112**, 4186 (2000).

[15] H. Ahlborn, X. Ji, B. Space, and P. B. Moore, J. Chem. Phys. **111**, 10622 (1999).

[16] T. Keyes and J. T. Fourkas, J. Chem. Phys. **112**, 287 (2000).

[17] B. Space, H. Rabitz, A. Lőrincz, and P.B. Moore, J. Chem. Phys. **105**, 9515 (1996).

[18] P. Moore and B. Space, J. Chem. Phys. **107**, 5635 (1997).

[19] P. B. Moore, X. Ji, H. Ahlborn, and B. Space, Chem. Phys. Let. **296**, 259 (1998).

[20] T. Keyes, J. Chem. Phys **106**, 46 (1997).

[21] J. R. Reimers and R. O. Watts, Chem. Phys. **91**, 201 (1984).

[22] M.C.C.Ribeiro, M. Wilson, and P. Madden, J. Chem. Phys. **110**, 4803 (1999).

[23] J. T. Kindt and C. A. Schmuttenmaer, J. Chem. Phys **106**, 4389 (1997).

[24] R. L. Murry, J. T. Fourkas, and T. Keyes, J. Chem. Phys **109**, 2814 (1999).

[25] R. B. Williams and R. F. Loring, J. Chem. Phys. **110**, 10899 (1999).

[26] S. Sastry, H. E. Stanley, and F. Sciortino, J. Chem. Phys. **100**, 5361 (1993).

[27] H. Ahlborn, X. Ji, B. Space, and P. B. Moore, J. Chem. Phys. **111**, 10622 (1999).

[28] X. Ji, H. Ahlborn, B. Space, P.B. Moore, Y. Zhou, S. Constantine, and L. D. Ziegler, Chem. Phys. **112**, 4186 (2000).

[29] R. Kubo, in *Fluctuation, Relaxation and Resonance in Magnetic Systems*, edited by D. T. Haar (Oliver and Boyd, Edinburgh and London, 1961).

[30] E. F. David and R. Stratt, J. Chem. Phys. **109**, 1375 (1998).

[31] R. A. Farrer, B. J. Loughnane, L. A. Deschenes, and J. T. Fourkas, J. Chem. Phys **106**, 6901 (1997).

[32] T. Kalbfleisch and T. Keyes, J. Chem. Phys **108**, 7375 (1998).

[33] B. J. Loughnane, A. Scodinu, R. A. Farrer, and J. T. Fourkas, J. Chem. Phys **111**, 2686 (1999).

[34] S. Mukamel, *Principles of Nonlinear Optical Spectroscopy* (Oxford University Press, Oxford, 1995).

[35] S.-B. Zhu, S. Singh, and G. W. Robinson, *Water in Biology, Chemistry and Physics* (World Scientific, New Jersey, 1996).

[36] J. E. Bertie and H. H. Eysel, Applied Spectroscopy **39**, 392 (1985).

[37] J. E. Bertie and Z. Lan, Applied Spectroscopy **50**, 1047 (1996).

[38] J. E. Bertie, M. K. Ahmed, and H. H. Eysel, J. Phys. Chem **93**, 2210 (1989).

[39] J. E. Bertie, the spectroscopic data is available in a convenient format at www.ualberta.ca/ jbertie/jebhome.htm.

[40] P. Madden and R. W. Impey, Chem. Phys. Let. **123**, 502 (1986).

[41] A. Wallqvist and O. Teleman, Mol. Phys. **74**, 515 (1991).

[42] P. Ahlström, A. Wallqvist, S. Engström, and B. Jönssom, Mol. Phys. **68**, 563 (1989).

[43] B. D. Bursulaya and H. J. Kim, J. Chem. Phys. **109**, 4911 (1998).

[44] M. Cho *et al.*, J. Chem. Phys. **100**, 6672 (1994).

[45] P. L. Silvestrelli, M. Bernasconi, and M. Parrinello, Chem. Phys. Lett. **277**, 478 (1997).

[46] M. Souaille and J. C. Smith, Mol. Phys. **87**, 1333 (1996).

[47] J. S. Bader and B. J. Berne, J. Chem. Phys. **100**, 8359 (1994).

[48] P. H. Berens and K. R. Wilson, J. Chem. Phys. **74**, 4872 (1981).

[49] D. Gragson and G. Richmond, J. Chem. Phys. **107**, 9687 (1997).

[50] D. Zimdars, J. Dadap, K. B. Eisenthal, and T. Heinz, J. Phys. Chem. B **103**, 3425 (1999).

Chapter 4

Low Frequency, Collective Solvent Dynamics Probed with Time-Resolved THz Spectroscopy

M. C. Beard, G. M. Turner, and C. A. Schmuttenmaer[*]

Department of Chemistry, Yale University, 225 Prospect Street, New Haven, CT 06520–8107

The time evolution of the change in the low-frequency response function for a solvent (*i.e.*, frequency-dependent absorption coefficient and index of refraction) following photoexcitation is experimentally determined for the first time. Unlike typical time resolved laser studies, the spectrum of the low frequency collective solvent modes are monitored rather than properites of the solute. We have investigated the dynamics of chloroform in response to photoexcitation of TBNC dye dissolved therein. We present frequency domain representations of this evolution with roughly 200 fs temporal resolution.

Introduction

Solvation and liquid dynamics play a very important role in essentially all liquid phase processes. Solvent effects are significant in photoexcitation, photoionization, electron transfer, proton transfer, energy dissipation, and liquid phase reactions. For example, the intermolecular solvent modes determine the ability of reactants to approach each other and products to separate from each other. It seems reasonable, therefore, to probe the low frequency solvent modes directly while one of the above processes is proceeding as a means to understand the solvent response to the event.

The general idea for the study reported here is that when a strongly absorbing chromophore is photoexcited, the surrounding solvent molecules will rearrange and adjust to this new charge distribution. Indeed, there have been many studies that monitor the fluorescence of the chromophore as a function of time after photoexcitation.[1] While those experiments reveal the timescale of solvent reorganization, the modes that are most strongly coupled to the solvation process are not explicitly probed. Molecules whose dipole moment dramatically increases or decreases upon photoexcitation are most often used for this type of study, but even non-dipolar solutes can be used because a change in the local charge distribution near a solute molecule can induce changes in the surrounding solvent modes.

TBNC (2,11,20,29-tetra-*tert*-butyl-2,3-naphthalocyanine), whose structure is shown in Fig. 1, was chosen as the dye molecule because it has a very small dipole moment in its ground and excited electronic states. Therefore, the dynamics probed are due to solute-solvent interactions and/or solvent-solvent interactions, while the pure solute response will be negligibly small. It is worth noting the importance of phthalocynanines, the class of molecules on which TBNC is based. They have been used extensively as dyes since their discovery.[2] Furthermore, they are the active material in Xerography and in laser printers. More recently, they have shown promise in photodynamic cancer therapy: they absorb in the red region of the visible spectrum, where light penetrates tissue best, they aggregate near tumors, and they have low fluorescence quantum yield, allowing a large amount of reactive singlet oxygen to be generated selectively near the tumor. They also appear to be useful materials for nanoelectronics and molecular electronics. While this report focuses on the solvent response, it would be interesting to carry out a study of the solute

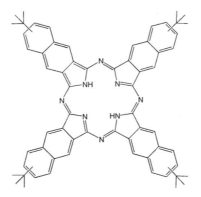

Figure 1. Structure of TBNC.

dynamics in the future utilizing optical techniques (800 nm pump/continuum probe) to directly compare the solvent and solute dynamics.

Studies of a wide variety of similar molecules based on TBNC will be instrumental in separating solute effects from solvent effects. That is, metal ions such as Cu^{2+}, Fe^{2+}, Zn^{2+}, or Vn^{2+} can be incorporated into the central cavity, which allows metal to ligand charge transfer (MLCT) and ligand to metal charge transfer (LMCT) excitation. A variety of functional groups can be attached peripherally, either in addition to, or instead of the *t*-butyl groups, thereby leading to differing interactions with the solvent. Furthermore, TBNC is soluble in a variety of solvents which will allow the dynamics of many systems to be measured and compared.

The generation of FIR (THz) pulses with sub-picosecond duration is a relatively new development,[3] and the full capability of this technology has only begun to be tapped. The spectral content of these pulses extends from 3 to 100 cm^{-1} (0.09 to 3 THz). Because THz pulses are generated from ~100 fs visible pulses, they can easily be used in a visible pump – THz probe configuration. Thus, low frequency intermolecular dynamics and solvation in liquids can now be studied directly in the time domain with previously unattainable temporal resolution.[4,5] One of the significant advantages of THz pulse spectroscopy is that the transient electric field itself is measured rather than its intensity. This allows full characterization of the complex-valued dielectric constant without having to carry out a Kramers-Kronig analysis.

There have been many papers that utilize the broadband FIR coverage of THz pulses, but only a few that do so in a time-resolved fashion. Of these, there have been 3 studies of the average change in THz transmission of a solution upon photoexcitation of a dye molecule.[4,5] This type of study reveals the overall timescale of the photoinduced change, but does not provide information regarding the solvent modes that are most active during solvation. In the present study, we extract the time-dependent change in intermolecular response function through use of time-resolved THz spectroscopy (TRTS). As described in previous work,[6,7] the measured change in THz transmission depends on two time variables, and the only way to obtain the time-dependent response function is to perform a series of scans and build up a 2-dimensional (2D) data set.

Experimental Methods

A 1 kHz regeneratively amplified Ti:Sapphire laser (Spectra Physics) is used to generate and detect the fs-THz pulses, and also serves as the photoexcitation source, as shown in Fig. 2. The THz pulses are generated by optical rectification of the visible pulse with a 1.0 mm thick ZnTe (110) single crystal,[8] and are detected *via* free space electro-optic sampling (FSEOS) in a separate 1.0 mm thick ZnTe (110) single crystal.[9] The THz pulse amplitude is typically 1000 times greater than the baseline noise, as measured with lockin detection with a 10 ms time constant. A typical THz waveform is shown in Fig. 2a. Its power spectrum, shown in Fig. 2b, extends from about 6 cm^{-1} to 80 cm^{-1}. About 60 µJ/pulse is used to photoexcite the dye molecules within a beam waist of roughly 2 mm. The 2D data sets are collected using a 30 ms lockin time constant, and require 8 seconds to acquire a 1024 point scan. Signal averaging of about 25 scans per pump-probe delay time is carried out to improve the signal to noise ratio (SNR), and about 30 pump-probe delay times are included in the 2D set.

The beam waist for the pump pulse is about 1.5 to 2 times smaller than that for the THz probe beam. We utilize a 1 mm diameter metal aperture to ensure that the frequency components of the THz beam that reach the detector have all passed through the photoexcited region, thereby minimizing distortion of the spectral changes.

Optical delay lines allow either the arrival time of the pump pulse to be varied while monitoring one point on the THz waveform, or measurement of the THz waveform at various fixed pump-probe delay times. It is often best to simultaneously scan the pump and probe delay lines in order to ensure that any

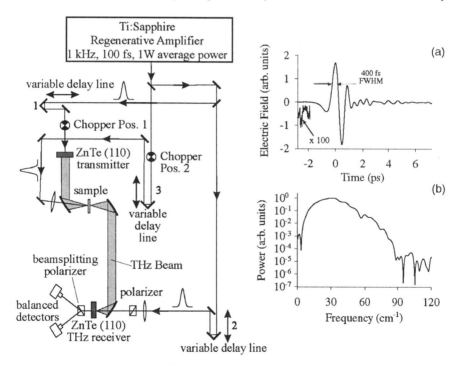

Figure 2. The experimental arrangement. Insets (a) and (b) present the THz pulse in the time and frequency domains, respectively. Reproduced with permission from reference 7. © American Physical Society 2000.

given point in the THz waveform samples the medium at a specific time after photoexcitation. This is most easily accomplished by adding a delay line for the transmitter pulse and scanning it in the opposite direction.[6,7]

Concentrations of TBNC in various solvents ranging from 3 to 30 mM have been investigated, and the dynamics are not concentration dependent; typically 10 mM is used. The TBNC molecules do in fact dimerize, but the absorption maximum of the dimers is shifted to shorter wavelengths to such a degree that the 800 nm pump pulse is primarily absorbed by monomers. The dye solution is flowed through a 200 μm thick cuvette at a rate such that each molecule experiences roughly 5 laser pulses as it passes through the beam waist, and we have verified that our signal is not dependent on flow rate.

In the strictest sense, the temporal resolution of this method is not determined by the duration of the THz pulse, but rather by the response time of the detector and dispersion in the sample. We estimate that our pump-probe temporal resolution is better than 200 fs.

Results and Discussion

General Considerations

TBNC has roughly D_{2h} symmetry with an absorption band at 800 nm, which is analogous to the Q band in porphyrins. This band is comprised of a pair of nearly degenerate $\pi \rightarrow \pi^*$ transitions that are not resolved. The ground state in the analogous porphyrin has a_u symmetry, and the nearly degenerate upper states have b_{2g} and b_{3g} symmetry. The z axis is defined as the symmetry axis. Both x and y transitions are allowed, where x and y are the axes within the molecular plane. The fluorescence quantum yield is estimated to be about 0.05, based on comparisons with phthalocyanines.

We find that the TBNC/chloroform system provides a reasonably good SNR because the chloroform solution does not absorb strongly in the FIR (the Naperian power absorption coefficient peaks at roughly 20 per cm), yet the chloroform molecule possesses a reasonably large dipole moment.

Before presenting the actual data, it is worth discussing a few issues related to the fact that the response function of the dye/solvent system is changing on a timescale that is fast compared to the THz pulse itself.[6] When a THz pulse propagates through a liquid under equilibrium or steady state conditions, the transmitted pulse can be described by convolving the input pulse with a response function that is based on the susceptibility of the medium.

$$E_{out}(t) = \int_{-\infty}^{\infty} \tilde{E}_{in}(\omega) \, \exp\left[-i \, \frac{\omega}{c} l \, \hat{n}(\omega)\right] \exp(i\omega t) \, d\omega \,, \qquad (1)$$

where $\tilde{E}_{in}(\omega)$ is the frequency-domain representation of the input pulse obtained by Fourier transformation, and $\hat{n}(\omega)$ is the complex-valued index of refraction of the medium, with $\hat{n}(\omega) = n(\omega) - ik(\omega)$, where n is the index of refraction, and k is related to the absorption coefficient through $k = \alpha c/(2\omega)$

In the laboratory we measure $E_{in}(t)$ and $E_{out}(t)$, but the susceptibility is the quantity of fundamental importance. It contains the underlying physics of the system under study. The susceptibility is the frequency domain representation of the time domain response function. The response function acts as a memory function, or kernel, to govern the polarization induced in the medium by an applied electric field:

$$P(t) = \varepsilon_0 \int_{-\infty}^{t} E(t') \chi(t - t') dt' \,. \qquad (2)$$

The susceptibility is related to the complex-valued index of refraction through the relations $1 + \hat{\chi}(\omega) = \hat{\varepsilon}(\omega) = \hat{n}^2(\omega)$ where $\hat{\varepsilon}(\omega)$ is the complex dielectric constant. It is possible to determine the susceptibility by deconvolving the input pulse from the transmitted pulse, as seen in eq 1. However, it is well known that deconvolution procedures can be adversely affected by experimental noise, and sometimes it is preferable to numerically propagate the THz pulse through the sample in order to determine the time-domain response function.[10]

The *change* in the response function, as is determined in a time-resolved experiment, can be rather small. Thus, noise introduced in numerical deconvolution procedure poses an even greater problem than in steady-state experiments. Furthermore, a response function with two time differences (the pump pulse arrival time, and the THz probe pulse time) is more complicated than that shown in eq 2. Therefore, we model our experiments in the time-domain through numerical propagation of the input THz pulse employing a parameterized response function.[11] As previously described,[6,7] we account for both time variables in a forward propagation. We determine the change in χ by implementing a nonlinear least squares fit of the model to the measured data.

Model Used

There are two contributions to the observed signal, a purely electronic part and a molecular part. The electronic part is instantaneous, and only active during the pump pulse duration. The molecular contribution consists of the solvent molecules responding to a new charge distribution, and its duration is a measure of the timescale of solvation. Only 2 to 5% of the solvent molecules participate in solvation and it is these molecules that are perturbed by photoexcitation. Therefore, we model the experiment with three response functions, one for the electronic response, and two (perturbed and unperturbed) for the molecular response, and these contributions depend on the pump-probe delay time t''. Thus, the overall time-dependent response function at time t and position z, given that the sample was photoexcited at time t'' is:

$$\chi(t,t'',z) = f_e(t,t'',z)\chi_e(t) + (1 - f_e(t,t'',z))\chi_0(t) + a(t,t'',z)\chi_{elc}(t) , \quad (3)$$

where $f_e(t,t'',z)$ is the "population" of oscillators affected by the photoexcitation nearby at time $t - t''$. The fraction not affected by photoexcitation is simply $1 - f_e(t,t'',z)$. The contribution of the instantaneous electronic response is given by $a(t,t'',z)$. The excited, non-excited, and electronic response functions are $\chi_e(t)$, $\chi_0(t)$, and $\chi_{elc}(t)$, respectively.

The response functions are treated as Lorentzian oscillators. The Lorentzian oscillator model describes a damped, driven harmonic oscillator. The equation of motion for the "position" of an oscillator as a function of time is

$$\ddot{s}_j(t) + \omega_j^2 s_j(t) + 2\gamma_j \dot{s}_j(t) = \sqrt{\frac{\eta_j \varepsilon_0}{m_j^* V}} E(t) , \quad (4)$$

where s_j is the macroscopic mode coordinate that represents the expectation value over all microscopic orientations and phases of the jth mode, ω_j is the jth oscillator frequency, γ_j is the jth damping coefficient, η_j is the coupling of the jth mode to the driving field $E(t)$, m^* is the effective mass of the oscillator, and V is the volume. The coupling to the field can be expressed as $\eta_j = \omega_j^2 (\varepsilon_s - \varepsilon_\infty) G_j$, where ε_s and ε_∞ are the static and high frequency dielectric constants, respectively, and $G_j = (\varepsilon_j - \varepsilon_{j+1})/(\varepsilon_0 - \varepsilon_\infty)$ if there is more than one mode (pole). The mode is overdamped if $\gamma_j > \omega_j$, and underdamped if $\gamma_j < \omega_j$. The time domain response function is the impulsive solution to eq 4.[11]

The temporal dependence of the excited oscillators is given by the convolution of the visible pulse (taken as a Gaussian function) and a single exponential decay, with time constant τ. The spatial dependence as a function of distance into the sample is a single exponential function determined from the visible absorption coefficient of the dye solution.

We numerically propagate a THz pulse through a medium that has the response function given by eq 3 by solving Maxwell's equations with a finite-difference time-domain (FDTD) procedure, as explained in detail in reference

11. It is desirable to carry out the propagation solely in the time-domain for several reasons. For example, complications arise because the absorption coefficient and refractive index of the medium change as the THz pulse propagates through it since τ is on the order of the THz pulse duration. Thus, different portions of the THz pulse will experience different optical parameters during propagation. Furthermore, the slowly varying envelope approximation is not valid for the THz pulses. Finally, mismatch between the group velocity of the visible pulse and the phase velocities of the various THz frequency components are easily accounted for, as are issues related to multiple reflections at the interfaces.

Transient FIR absorption by the dye molecule itself has been considered and ruled out for 3 reasons. First, while a low level Gaussian[12] calculation (RHF/STO-3G) identified 12 vibrational modes below 100 cm^{-1}, their IR and Raman intensities are over 1000 times weaker than the strongest peaks (which occur near 1200 and 2000 cm^{-1}), and all but two are actually 10^5 times weaker than the strongest peaks. Second, the difference spectra observed are much broader than one would expect for an intramolecular mode. Third, a degenerate pump/probe experiment at 800 nm, which is sensitive to the response of the dye molecules, was carried out and shows a completely different time dependence than the low frequency transients (Figure 3).

Results of Fit

We employ a non-linear least squares fitting routine to adjust the parameters in our model to obtain the best fit. The results of the FDTD-TRTS fit are shown in Fig 4. Figure 4a displays a contour plot of the measured data, and Fig. 4b is the best fit calculated data using the parameters given in Table 1. Figures 4c and 4d are representative cuts at pump-probe delay times of 0.65 and

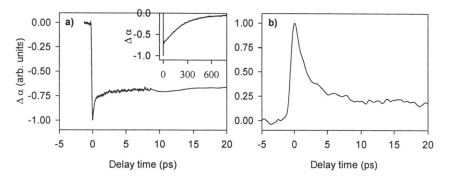

Figure 3. Comparison of time dependence of degenerate 800 nm pump/probe experiement (part a) and TRTS experiment (part b).

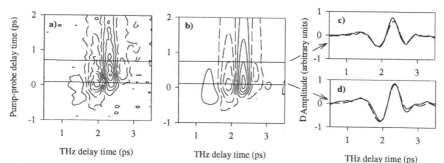

Figure 4. Contour plot of 2D TRTS difference signal (THz pulse propagated through sample with laser on minus that with laser off) for TBNC/chloroform. Solid contours correspond to positive values, and dashed contours are used for negative ones. Comparison of measured (part a), with calculated (part b) values. Parts c) and d) display two representative cuts through the surfaces as indicated by the horizontal lines in parts a) and b). In parts c) and d), the measured data are shown with solid lines, and results of the numerical propagation are shown with the dashed lines.

0.05 ps, respectively, which compare the calculated and measured data. We collect the data as described in Refs. 6 and 7 such that every portion of the probe has experienced the same delay from the pump beam. The ground state spectrum of chloroform is well fit by a two-pole Lorenztian function (see Fig. 5) and its parameters are held fixed during the fit. During the simulations, the width of the Gaussian excitation pulse is held fixed at 150 fs, a value slightly greater than that determined from the autocorrelation (110 fs) because of slightly noncollinear propagation of the THz and visible pulses through the sample and/or the THz detector crystal.

Figure 6 shows the frequency-dependent absorption coefficient and index of refraction as a function of time during and after photoexcitation. The peak of the absorption coefficient increases after photoexcitation. In addition, there is an electronic contribution at low frequencies as manifested in the sharp feature that only persists as long as the excitation pulsewidth. The refractive index shows a fairly uniform increase, but with a slightly greater increase at low frequencies. There is also an electronic contribution at low frequencies at time $t = 0$.

In Figure 7, the *change* in $\alpha(\omega)$ and $n(\omega)$ is plotted instead. One can more easily see which spectral regions are undergoing changes. At low frequencies, the index of refraction initially gets smaller before increasing. That is, the instantaneous electronic contribution is negative, but the molecular contribution is positive. The return of the absorption coefficient and refractive index back to their steady state values occurs with a time constant of $\tau = 1.40$ ps. The excited

**Table 1. Lorenztian oscillator parameters for TBNC/chloroform. Asterisks
denote the photoexcited medium.**

Oscillator #	η_i (cm^{-1})^2	η_i^* (cm^{-1})^2	ω_i (cm^{-1})	ω_i^* (cm^{-1})	γ_i (cm^{-1})	γ_i^* (cm^{-1})
1	171.4	1700(20)	41.60	97.9(.1)	30.56	174.9(.1)
2	115.3	180(5)	6.57	8.3(.2)	19.35	29.5(.1)
electronic	---	4.3(.2)	---	2.4(.1)	---	6.5(.1)

Other Parameters

lifetime (τ)	1.40(0.01) ps
visible pulsewidth (Δw)	150 fs (held fixed)
fraction excited (f_e)	0.05 (held fixed)
optical skin depth (δ)	64 μm (held fixed)

state lifetime of the TBNC dye molecule is on the order of 600 ps, which
implies that τ corresponds to the timescale of solvent rearrangement rather than
a solute mode.

As seen in Figure 5 and Table 1, the mode at 6.57 cm^{-1} is overdamped
(non-resonant), and the mode at 41 cm^{-1} is slightly underdamped (resonant) in
the FIR static spectrum of CHCl$_3$. The 6.57 cm^{-1} mode can be associated with a
Debye response with a corresponding Debye time of about 4.6 ps. The
overdamped Lorentzian function simply allows the absorption to decrease,
thereby avoiding the Debye plateau. The slightly resonant 41 cm^{-1} mode
accounts for the well known "Poley" absorption found in dipolar fluids.[13] This

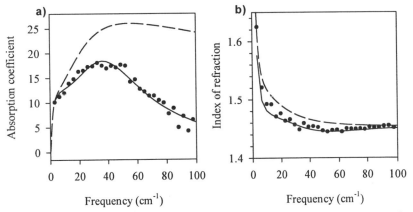

*Figure 5. Far-IR optical properties of CHCl$_3$. Points are experimental
results, solid line is fit based on 2-pole Lorenztian oscillator, and the dashed
line represents the medium after photoexcitation of the dye.*

a)

b)

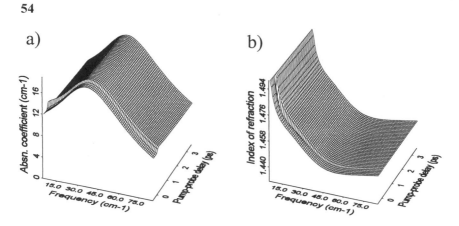

Figure 6. Time-dependent FIR absorption coefficient (part a) and refractive index (part b) of CHCl₃ after photoexcitation of TBNC.

mode can mostly be associated with librational motion of the $CHCl_3$ molecules, along with some collision-induced dipole interactions. The absorption coefficient for the excited state and ground state oscillators are compared in Fig. 5. Upon photoexcitation, the Lorenztian oscillators lose the resonant character of the librational mode. That is, the mode changes from underdamped to overdamped, which is also represented in the best-fit parameters (see Table 1). The loss of librational character is due to a molecular reorientation immediately after photoexcitation. That is, they are reorienting rather than librating at this time. The equilibrium librational period is 0.8 ps, compared to the solvation time of 1.4 ps. The fact that these timescales are so similar, and the loss of librational resonance suggest a highly transitory state for the $CHCl_3$ molecules that are involved in solvation. Finally, the fact that the mode shifts to higher frequencies during solvation can not be associated with an increase in local structure because of the simultaneous loss of resonant character.

The Debye process is a single molecule response and is described by the single molecule dipole correlation function. It represents the rotational diffusion of a single dipole in a dielectric medium but excludes any dipole-dipole interactions, that this, the pure rotational gas phase energy levels are extremely broadened in solution through collisions. The slight increase in ω_i for oscillator #2 suggest that the solvating molecules are undergoing a smaller number of collisions, or that they are rotating faster. The rapid change in their electrostatic environment may give them an additional 'kick'. The fact that solvation occurs faster that the Debye time suggests that the electrostatic interactions, embodied in the librational motion, are more important than the collisional interactions.

The dipole moment of TBNC changes very little upon photoexcitation. There is some charge redistribution, as shown in Fig. 8a, but the overall electrostatic properties of the excited state are not very different from that in the ground state. This is consistent with the fact that the position of the Q band in

a)   b)

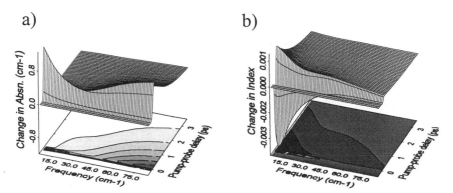

Figure 7. Time-dependent change in absorption coefficient (part a) and refractive index (part b) of CHCl₃ after photoexcitation of TBNC.

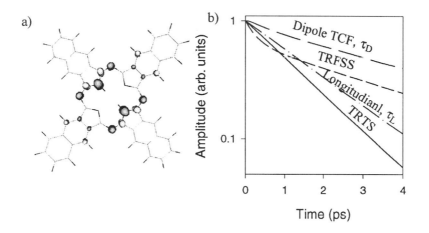

Figure 8. Part a) shows the change in charge distribution of TBNC upon photoexcitation. The dark features indicate a gain of electron density, and the lighter features indicate a loss. The largest changes correspond to roughly 1% of an electron charge. Part b) is a semilog plot of various relaxation times as indicated on the figure and discussed in the text.

TBNC is essentially independent of solvent polarity. The solvent molecules initially respond to the new charge distribution, but once they have rearranged themselves, their environment is much as it was prior to photoexcitation. Thus, the solvent dynamics relax back to their non-photoexcited behavior after undergoing a transient deviation.

Figure 8b compares the TRTS lifetime with time-resolved fluorescence Stokes shift (TRFSS) and steady-state dielectric relaxation experiments. There is general agreement on the overall timescale for all of these results. Our results are shown with the solid line, the dipole time correlation function (TCF) is shown with the long dashed line,[14] and the solvation TCF from the TRFSS experiments is shown with the short dashed line.[15] The longitudinal response TCF is shown with the dot-dashed line. The dipole TCF has a relaxation time constant given by τ_D, the Debye relaxation time, and the longitudinal relaxation time, given by τ_L, is obtained from the Debye relaxation time using $\tau_L = \tau_D(\varepsilon_\infty/\varepsilon_0)$. Recent work has indicated that the solvation response from the TRFSS experiments should be nearly identical to the longitudinal dielectric response of the solvent.[16] However, it is seen that the fluorescence Stokes shift TCF is too long in comparison to the longitudinal TCF. In fact, it has roughly the same behavior as the Debye TCF rather than the longitudinal TCF. The relaxation behavior from TRTS experiments agrees semiquantitatively with these other techniques, and the deviations from the other techniques are of the same magnitude as the deviations of the TRFSS results with the relaxation time constants. Interestingly, the TRTS lifetime agrees most closely with the longitudinal TCF.

Conclusions

The importance of the TRTS results is that in addition to the overall relaxation times, we are able to monitor changes in the low frequency solvent spectrum. Upon photoexcitation of the dye molecule, the absorption peak is significantly broadened due to loss of the resonant character of the high frequency mode, and the change in its refractive index favors slightly higher values. This information is not available from dynamic Stokes shift experiments. Future work will compare the results of several solvents in an effort to uncover similarities and differences.

The data are most easily interpreted via numerical pulse propagation rather than Fourier deconvolution techniques. One reason for this is the finite signal to noise ratio of experimental data, but there are more fundamental reasons that are related to dispersion of the THz pulse in the sample as it propagates.

Acknowledgments

This work was partially supported by the National Science Foundation, and the Sloan Foundation.

References

[1] For reviews of the application of time-dependent Stokes shift experiments to solvation dynamics, see Maroncelli, M. *J. Mol. Liq.* **1993**, 57, 1; Barbara, P. F.; Jarzeba, W. *Adv. Photochem.* **1990,** 15, 1.

[2] McKeown, N. B. *Phthalocyanine Materials: Synthesis, Structure, and function;* Cambridge University Press: Cambridge, U.K., 1998.

[3] Fattinger, Ch.; Grischkowsky, D. *Appl. Phys. Lett.* **1989,** 54, 490; Smith, P. R.; Auston, D. H.; Nuss, M. C. *IEEE J. Quant. Electron.* **1988,** 24, 255.

[4] Haran, G.; Sun, W. D.; Wynne, K.; Hochstrasser, R. M. *Chem. Phys. Lett.* **1997,** 274, 365. McElroy, R.; Wynne, K. *Phys. Rev. Lett.* **1997,** 79, 3078.

[5] Venables, D. S.; Schmuttenmaer, C. A. In *Ultrafast Phenomena XI;* Elsaesser, T.; Fujimoto, J. G.; Wiersma, D.; Zinth, W., Eds.; Springer-Verlag: Berlin, 1998; pp. 565-567.

[6] Kindt, J. T.; Schmuttenmaer, C. A. *J. Chem. Phys.* **1999,** 110, 8589.

[7] Beard, M. C.; Turner, G. M.; Schmuttenmaer, C. A. *Phys. Rev. B,* **2000**, 62, 15764

[8] Rice, A.; Jin, Y.; Ma, F.; Zhang, X.-C. *Appl. Phys. Lett.* **1994,** 64, 1324.

[9] Lu, Z. G.; Campbell, P.; Zhang, X. -C. *Appl. Phys. Lett.* **1997,** 71, 593.

[10] Harde, H.; Keiding, S.; Grischkowsky, D. *Phys. Rev. Lett.* **1991,** 66, 1834.

[11] Beard, M. C.; Schmuttenmaer, C. A. *J. Chem. Phys.,***2001**, 114, 2903.

[12] Frisch, M. J. *et al;* Gaussian 95, Development Version (Revision E.1), Gaussian, Inc., Pittsburgh PA, 1996.

[13] Gerschel, A; Darmon, I; Brot, C. *Mol. Phys.* **1972,** 23, 317.

[14] Kindt, J. T.; Schmuttenmaer, C. A. *J. Chem. Phys.* **1997,** 106, 4389.

[15] Horng, M. L.; Gardecki, J. A.; Papazyan, A.; Maroncelli, M. *J. Phys. Chem.* **1995,** 99, 17311.

[16] Castner, E. W., Jr.; Maroncelli, M. *J. Mol. Liq.* **1998,** 77, 1.

Chapter 5

Translational Dynamics of Fluorescently Labeled Species by Fourier Imaging Correlation Spectroscopy

Michelle K. Knowles[1], Daciana Margineantu[2], Roderick A. Capaldi[2], and Andrew H. Marcus[1,*]

[1]Department of Chemistry and Materials Science Institute, University of Oregon, Eugene, OR 97403
[2]Department of Biology and Institute of Molecular Biology, University of Oregon, Eugene, OR 97403

Time-dependent spatial distributions of density fluctuations in synthetic and biological systems are determined from purely incoherent fluorescence signals. This is accomplished using a new method, Fourier imaging correlation spectroscopy (FICS), that is based on the detection of modulated fluorescence signals and measures temporal fluctuations of a spatial Fourier component of the sample particle number density. The information contained by FICS measurements provides details about the spatial relationship between fluorescent species, usually only obtained by direct imaging single-particle experiments. The FICS approach offers significant advantages in signal-to-noise detection efficiency, allowing a broader dynamic range to be accessed experimentally.

The dynamics of complex fluids is an area of fundamental importance and technological relevance. Macromolecular or mesoscopic translational motion is important in many different kinds of soft-matter systems ranging from self-organized block-copolymer films to protein transport in live biological cells. Traditionally, experimental information about the structure of complex fluids is obtained by light scattering from ordered arrangements of atoms, molecules or larger scattering centers (*1*). Dynamics are studied by performing dynamic light scattering (DLS) measurements of the fluctuations of scattered light intensity (*2*). Such experiments reveal the existence of multi-exponential relaxations that arise from the complex interactions between fluid components occurring over a range of spatial and temporal scales (*3*). Nevertheless, many intriguing soft materials have been left unexplored due to a lack of sufficient light scattering contrast. In this chapter we present an overview of a new method called Fourier imaging correlation spectroscopy (FICS) that retains the wave number selectivity of DLS while overcoming many of its limitations in sensitivity (*4,5*).

Fourier Imaging Correlation Spectroscopy

In FICS experiments, modulated fluorescence signals are detected from the intersection of an excitation fringe pattern (an optically generated grating) with a microscopically heterogeneous configuration of chromophores, $C(\mathbf{r},t)$. The basic principle is illustrated in Figure 1A for the case of a two-dimensional fluid. A system of N uniformly labeled fluorescent particles (shown as discs with diameter σ) is illuminated by an excitation grating having a fringe spacing d_G.

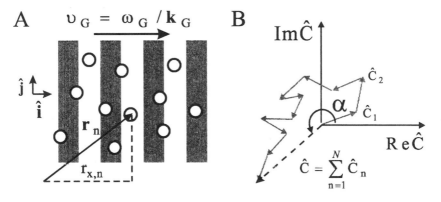

Figure 1. (A) Schematic of the FICS experimental geometry. Fluorescent particles are represented as circles and the excitation grating as gray bars. (B) A static particle configuration is uniquely described by a sequence of vectors in the complex plane whose superposition is $\hat{C}$.

The grating is characterized by a wavevector, $\mathbf{k}_G = 2\pi d_G^{-1}\hat{\mathbf{i}}$, that is directed along the x-axis. The time-dependent microscopic density is given by

$C(\mathbf{r}, t) = \frac{1}{V} \sum_{n=1}^{N} A_n \delta[\mathbf{r} - \mathbf{r}_n(t)]$, where A_n is the area of the nth particle, proportional to that particle s fluorescence intensity, and V is the total area of the illuminated system. The phase of the grating is modulated at the angular frequency ω_G so that its position is swept across the system at a velocity, $\upsilon_G = \omega_G / \mathbf{k}_G$, much greater than the average speed that a particle travels the interfringe distance. The resulting fluorescence intensity has the functional form (4,5),

$$I_G(t) = (\kappa/V) I_0 \left\{ \hat{C}(0) + \left| \hat{C}(\mathbf{k}_G, t) \right| \cos[\omega_G t + \alpha(\mathbf{k}_G, t)] \right\}, \qquad (1)$$

where κ is the proportionality factor that accounts for the absorption cross section, quantum yield, and light collection efficiency of the experimental setup, and I_0 is a constant intensity level. According to eq 1, the signal is composed of two parts; a stationary (or dc) component that represents the zero-k background fluorescence level, and a modulated (or ac) component whose amplitude, $|\hat{C}|$, and phase angle, α, depends on the particle configuration. The physical meaning of $|\hat{C}|$ and α are the time-dependent modulus and phase angle, respectively, associated with a single Fourier component, $\hat{C} = |\hat{C}| \exp(i\alpha)$, of the transformed microscopic density evaluated at the wave vector, $\mathbf{k}_G$:

$$\hat{C}(\mathbf{k}_G, t) = \left(\frac{1}{2\pi} \right)^{1/2} \left(\frac{1}{V} \right) \sum_{n=1}^{N} A_n \exp[i\mathbf{k}_G \cdot \mathbf{r}_n(t)] . \qquad (2)$$

Because particle motion occurs on a much slower time scale than the inverse modulation frequency, the lock-in detection method (5) is used to demodulate the signal into slowly varying complex components of $\hat{C}$, namely $\mathrm{Re}\hat{C} = |\hat{C}| \cos\alpha$ and $\mathrm{Im}\hat{C} = |\hat{C}| \sin\alpha$. As shown in Figure 1B, the experimentally detected Fourier component, $\hat{C}$, is a vector superposition of N microscopic particle terms. Each particle contributes an amplitude, A_n, and a phase angle, $\alpha_n = \mathbf{k}_G \cdot \mathbf{r}_n = k_G r_{x,n}$, to the detected value of $\hat{C}$. Since $\mathbf{k}_G$ points in the direction of the x-axis, only the x-components of the positions $\mathbf{r}_n$ contribute to the phase angles. For the case of a homogeneous isotropic fluid in the absence of external fields, measurements carried out at a single orientation of $\mathbf{k}_G$ is equivalent to all other orientations.

Fluctuations of the vector $\hat{C}$ in the complex plane results as a direct consequence of collective particle density fluctuations in real space. If the particle motion is random on the length scale of d_G, $\hat{C}$ is expected to follow a stochastic trajectory in the complex plane. The trajectory of $\hat{C}$ can be used to construct the probability distribution $P[\hat{C}]$, and therefore all of the time-correlation functions associated with $\hat{C}(\mathbf{k}_G,t)$.

Temporal and spatial two-point correlation functions of the fluid are determined directly from the trajectory of $\hat{C}(\mathbf{k}_G,t)$ according to

$$F(\mathbf{k},\tau)=\left\langle \hat{C}^*(\mathbf{k},t)\hat{C}(\mathbf{k},t+\tau)\right\rangle \text{ and } S(\mathbf{k}) = \left\langle \hat{C}^*(\mathbf{k},t)\hat{C}(\mathbf{k},t)\right\rangle. \qquad (3)$$

The function $F(\mathbf{k},\tau)$ is called the intermediate scattering function and is the fundamental quantity of interest to the theoretical description of liquid state dynamics (*1,2,3*). The static structure factor, $S(\mathbf{k})$, is simply the $\tau = 0$ limit of $F(\mathbf{k},\tau)$. For homogeneous isotropic fluids, the correlation functions defined by eq 3 depend only on the magnitude, $k - |\mathbf{k}|$. Thus, $F(k,\tau)$ and $S(k)$ are independent of the orientation of $\mathbf{k}$. Roughly speaking, the decay time of $F(k,\tau)$ at a particular value of k is given by $\tau_0 = [D_0 k^2]^{-1}$, which is the time required for an unhindered particle to diffuse (with diffusion coefficient D_0) the distance k^{-1}.

Instrumentation

We show in Figure 2 a schematic of the apparatus we have constructed to perform FICS measurements. The sample is held at the object plane of a fluorescence microscope and placed at the focus (spot size, $w \cong 100\ \mu m$) of two linearly polarized intersecting laser beams. The excitation source is the continuous wave frequency doubled output of a Spectra Physics Nd:YAG laser ($\lambda_{ex} = 532$ nm); its output power (measured just before sample incidence) is typically set to below 0.1 mW to minimize photodegradation during data acquisition. Depending on the desired temporal resolution, data acquisition may vary between a few seconds to 60 minutes. The two beams produce an intensity interference fringe pattern inside the sample, with adjustable spatial period d_G, according to $d_G = \lambda_{ex}/2n\sin(\varphi/2)$, where n is the relative refractive index of the dielectric interface upon which the beams are incident, and φ is the angle between the two beams (6).

One of the beams is passed through an electro-optic phase-modulator (Conoptics). A frequency generator (Keithley) is used to modulate the phase of the excitation grating from 0 to 2π at the angular frequency ω_G ($\cong 10 — 100$ KHz). The modulated fluorescence signal, $I_G(t)$, is collected using a fused silica oil-immersion objective (Leica, Plan Fluotar, 100×, Na = 1.3).

The emission collected by the objective is passed through a stack of glass absorption filters (Corning, cutoff wavelength 570 nm, transmission efficiency 70% at 605 nm), an interference band-pass filter (CVI Laser, central

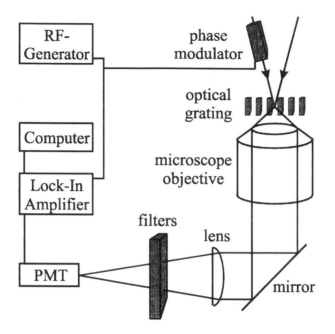

Figure 2. Schematic of the FICS apparatus.

wavelength 605 nm, bandwidth 10 nm, transmission efficiency 90%) and an excitation barrier filter. The filtered signal is imaged onto a thermoelectrically cooled photomultiplier tube (PMT, Hamamatsu, R1527) operating in photon-counting mode. The PMT output is detected using a digital dual-phase lock-in amplifier (Stanford Research Systems, SR830) that is referenced to the signal waveform used to drive the phase modulator. A computer, which controls an analog-to-digital data acquisition board (National Instruments), records separately the average background fluorescence intensity, $\kappa I_0 \hat{C}(0)$, the complex components of the demodulated signal, $\mathrm{Re}\hat{C}$ and $\mathrm{Im}\hat{C}$, and the laser excitation power.

In practice, the mean fluorescence intensity, after being corrected for drifts in laser power (less than $\pm$ 1%), is used to normalize the demodulated signal, effectively removing the influence of photodegradation. The ability to detect this signal is determined by the signal-to-background ratio defined by $S/B = \kappa I_0 \langle \hat{C}^2 \rangle / \kappa I_0 \hat{C}(0) = [\frac{1}{2} V\hat{C}(0)]^{-1/2}$, where V is the excitation volume. In our experiments on dense colloid suspensions (described below), the reduced areal density $\rho^* = N\sigma^2/A = 0.52$, $V \cong 9,400$ μm^3 and $C(0) \cong 0.45$ particles / μm^3 so that $S/B \sim 0.02$. Typically, 32,000 data points are collected at an acquisition frequency of 512 Hz. Individual data sets are repeated $\sim$ 10 times, cross-checked for consistency and averaged together. Under these conditions we found S / B ~ 2%.

Time-correlation functions of the $\hat{C}$ trajectories are computed by averaging over t_{max} time origins according to

$$\langle A(t)A(t+\tau)\rangle = (1/t_{max}) \sum_{t_0}^{t_{max}} A(t_0)A(t_0+\tau),$$ (4)

where $A = \mathrm{Re}\hat{C}, \mathrm{Im}\hat{C}$. The decay time of the autocorrelation function is a meaure of the time required for a fluorescent particle to move the distance k_G^{-1}. The upper limit to the temporal resolution of FICS measurements is determined by the modulation frequency. Essentially, 10 cycles of the reference oscillator are necessary to determine a single data point. For the experiments presented here, a modulation frequency of 50 KHz is used, corresponding to an instrumental time resolution of 200 µsec.

The FICS apparatus can also be run in direct visualization mode. In this way, microscopic information is obtained by recording sequenced images of the sample plane via digital video fluorescence microscopy (DVFM). This allows direct comparison to be made between the time-correlation functions measured by the FICS method and the same quantities calculated from the microscopic particle trajectories via statistical mechanics (5,7). The experimental procedure to perform DVFM measurements is described in detail by Crocker and Grier (8), who developed the original particle tracking algorithms.

Dynamics of Monolayer Colloid Suspensions

In Figure 3A is shown the results of FICS measurements performed on a dilute monolayer suspension of Rhodamine labeled poly(styrene) spheres (diameter, $\sigma = 1$ µm) with $\rho^* = 0.02$ ($N \approx 160$ particles contained by an area $A \approx 7850$ µm^2). At low density, the mean inter-particle separation is large, [$L = (\rho^*/\sigma^2)^{-1/2} \sim 7.1$ µm], such that the system behaves like a superposition of non-interacting Brownian particles for $d_G < L$. Under these conditions, the dynamic structure function takes the Gaussian form (2)

$$F_S(k,\tau) = \exp\left[-k^2 D_S \tau\right]$$ (5)

where $F_S(k,\tau)$ is called the self dynamic structure function and D_S is the self-diffusion coefficient. The data shown in Figure 3A decays in time as a single exponential and scales with wave-number as a Gaussian, in precise agreement with the theoretical prediction. The solid lines correspond to eq 5 with $D_S = 3 \times$

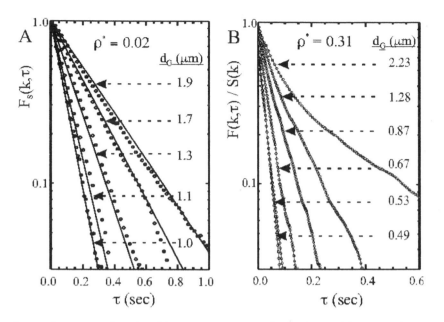

Figure 3. (A) Results of FICS measurements performed on a dilute monolayer colloid suspension and (B) a semi-dilute suspension. (Reproduced from reference 5. Copyright American Institute of Physics.)

10^{-9} cm^2 sec^{-1}. This value for the self-diffusion coefficient is in good agreement with that of the free diffusion coefficient, $D_0 = 0.707k_BT / 6\pi\eta a = 3.1 \times 10^{-9}$ cm^2 sec^{-1}, calculated from the Stokes-Einstein equation with a correction to account for the hydrodynamic friction due to the effect of the cell walls (5,7).

In Figure 3B is shown plots of $F(k,\tau)/S(k)$ as a function of time for a semidilute monolayer poly(styrene) suspension ($\rho^* = 0.31$). For this system, the dynamics is complicated by multi-exponential decays that vary with k. The mean inter-particle separation at this density is $L = 1.8$ μm. The data indicates that the particle dynamics exhibit multi-exponential behavior when the system is probed at $d_G = 2.23$ μm $> L$, but remains single-exponential for all fringe spacings $d_G = 0.49, 0.53, 0.67, 0.87, 1.28$ μm $< L$. For $d_G = 2.23$ μm, there is an apparent transition of the effective diffusion coefficient from short- to long-time behavior, which occurs at $\tau = 150$ msec. When the system is probed on length scales less than L, particles appear to diffuse freely without hindrance from neighboring particles. When the system is probed on length scales greater than L, the aparent diffusion coefficient is dressed by collisions between nearest neighbors, effectively decreasing the diffusion coefficient. The observed time scale of this kinetic transition is slightly smaller than the average collision time between particles ($\tau_c = [(2\pi / L)^2 D_0]^{-1} \cong 250$ msec).

When measurements are performed on dense crowded systems, the wave number and time-dependence of the dynamics becomes more complex. Figure 4A displays a direct comparison of $F(k,\tau)/S(k)$ determined from FICS data (solid curves) and from DVFM data (open circles) as a function of time for a monolayer poly(styrene) suspension with $\rho^* = 0.51$. The inset shows a fluorescence micrograph of a static particle configuration. Decays are shown for three different wave numbers. The FICS data were constructed according to eq 3, while the microscopy data were calculated by Fourier inversion of the microscopic particle trajectories as described in reference 6. Agreement between the two independent measurements of the intermediate scattering function is excellent. For the highest value of k shown ($d_G = 0.86$ μm), the decay is nearly single-exponential. As the wave number is decreased (or the fringe spacing is increased) the decays begin to exhibit multi-exponential character. Our interpretation of this result is that the high-k measurement is primarily sensitive to pre-collisional free motion, while the low-k measurements are sensitive to particle-particle interactions. An analogous comparison is shown for S(k), where the FICS data (filled circles) have been scaled by an arbitrary factor along the vertical axis. The agreement to microscopy data (solid curve) is very good, indicating that the FICS method is capable of measuring structural correlations of the fluid. In this example, the temporal resolution of the FICS measurement is

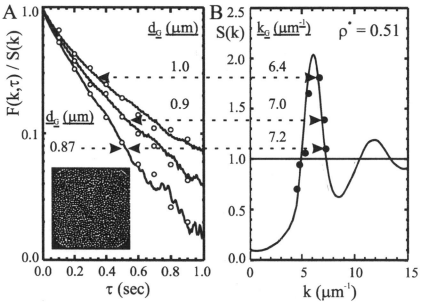

Figure 4. Comparison between FICS and DVFM measurements for a dense monolayer colloid suspension. Plots are shown for (A) the intermediate scattering function, and (B) the static structure factor. (Reproduced from reference 5. Copyright 2000 American Institute of Physics.)

2.0 msec, 5 times faster than current digital video technology. We expect future variations of the technique to access microsecond time scales.

Dynamics of Mitochondria in Live Cells

The ability to determine $F(k,\tau)/S(k)$ from weakly fluorescent biological samples is demonstrated in Figure 5. FICS experiments were performed on live human osteosarcoma cells (143B) treated with aqueous JC-1 solution (Molecular Probes, 0.25 μM, 5 minute exposure), a fluorescent dye ($\lambda_{em} = 590$ nm) that accumulates inside the mitochondrial compartment (9). The cells were cultured using HG-DMEM medium supplemented with 10% fetal calf serum, and incubated in 5% CO_2 atmosphere. The inset shows a contrast enhanced fluorescence micrograph, taken with an intensified CCD camera (Princeton Instruments Pentamax), of a similarly treated cell. In the samples we studied, the mitochondrion exists as a network of flexible filaments that constantly undergo rearrangements in position. Little is known about the details of this motion.

The average power level used during a ~45 minute data run was 0.1 mW. No laser induced pathological effects in cell behavior was observed. Figure 5A shows that the dynamics of the mitochondrion is complicated by multi-exponential relaxations; a full analysis is presented in reference 9. Comparison between FICS (solid curves) and microscopy (open circles) measurements of $F(k,\tau)/S(k)$ show excellent agreement between the two methods.

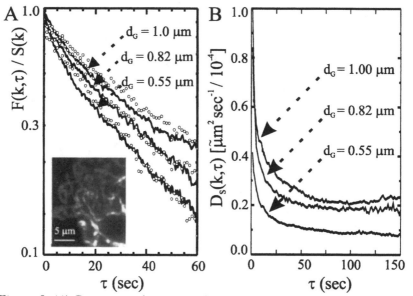

Figure 5. (A) Comparison between FICS and microscopy data for reticulated mitochondria in live cells. (B) The effective self-diffusion is plotted as a function of time. (Reproduced from reference 9. Copyright 2000 Biophysical Society.)

The kinetic behavior of mitochondrial filaments can be studied through the time- and wave number-dependent self-diffusion coefficient, $\tilde{D}_S(k,\tau)$. We calculate $\tilde{D}_S(k,\tau)$ from our FICS data according to $\tilde{D}_S(k,\tau) = -\ln[F_S(k,\tau)/k^2\tau]$, a generalization of eq 5. In Figure 5B are shown plots of $\tilde{D}_S(k,\tau)$ corresponding to the data shown in Figure 5A. For intermediate times (1 sec $< \tau <$ 60 sec) and fixed k, $\tilde{D}_S(k,\tau)$ decays on the time scale $\tau_l \sim$ 15 sec. This relaxation indicates a kinetic transition from short-time filament motion to a dressed collective long-time behavior. The values for $\tilde{D}_S(k,\tau)$ lie in the range 3.5 - 0.5 $\times$ 10^{-12} cm^2 sec^{-1}, consistent with velocities observed for mitochondria undergoing cytoskeletal-assisted directed motion ($\sim$50 nm sec^{-1}). Examination of the k-dependence of $\tilde{D}_S(k,\tau)$ at fixed τ reveals that the effective diffusion coefficient at all times is consistently smaller for $d_G \sim$ 0.55 μm than it is for $d_G \sim$ 0.82, 1.0 μm. Our observations suggest that the kinetic transition from short- to long-time behavior is the result of a structural rearrangement of the local mitochondrial filament environment on the length scale $d_l \sim$ 0.8 μm.

To examine the effects of metabolic activity we used FICS to study JC-1 labeled cells after incubation with drugs known to alter metabolism. In Plate 1 are shown fluorescence micrographs of JC-1 labeled cells after exposure to various drugs. JC-1 is a positively charged carbocyanine dye that is a quantitative fluorescence indicator of membrane potential, $\Delta\Psi$. Local regions of the membrane that are energized promote an uptake of JC-1 into the mitochondrial matrix with subsequent formation of a J-aggregate of JC-1 that emits yellow fluorescence ($\sim$590 nm). The monomer form of JC-1 emits green fluorescence. As shown in Plate 1A, the spatial distribution of $\Delta\Psi$ in reticulate mitochondria appears heterogeneous under physiological conditions. This heterogeneity is sensitive to the metabolic state of the cell. In Plate 1B, we show the effects of treatment with Nigericin, an ionophore that exchanges K$^+$ and H$^+$ across the mitochondrial inner membrane resulting in uncoupling of respiration from ATP production. The net effect of Nigericin treatment is the hyperpolarization of the mitochondrial inner membrane. The spatial distribution of $\Delta\Psi$ becomes uniformly large throughout the reticulum after $\sim$30 minutes incubation with Nigericin. In Plate 1C, we show the effects of inhibition of respiration. Antimycin A inhibits the activity of mitochondrial respiratory chain complex III. Cells treated with Antimycin A ($\sim$10 minutes incubation) show a progressive decrease in local membrane regions with high $\Delta\Psi$. In Plate 1D we show the effects of Staurosporine, a protein kinase inhibitor that induces apoptosis (programmed cell death). Dramatic changes in mitochondrial membrane morphology are observed in cells that have been treated with Staurosporine ($\sim$4 hours incubation). Staurosporine has the initial effects of hyperpolarizing the mitochondrial membrane, membrane swelling, and the disruption of the reticulum structure with the formation of giant mitochondrial vescicles.

We used FICS to study the motion of hyperpolarized regions of the mitochondrial membrane by detecting emission only at 590 $\pm$ 5 nm. In Figure 6, is shown direct comparisons between $\tilde{D}_S(k,\tau) \cdot \tau = -\ln[F(k,\tau)/k^2]$ for control cells and for those treated with the drugs described above. For these measurements the fringe spacing was set to $d_G = 0.8$ μm, the length scale associated with filament reorganization.

In Figure 6A are shown results for Nigericin treated cells in which mitochondrial ATP synthesis has been uncoupled from respiration. For short times ($\tau < 15$ sec) $\tilde{D}_S(k,\tau)\tau$ is indistinguishable from the corresponding control cell measurement. For long times ($\tau > 15$ sec), the slope, $\tilde{D}_S(k,\tau)$, is a factor of 1.5 smaller for Nigericin in comparison to control cells. The transition time ($\sim$ 15 sec) is the same as the interaction time scale, τ_I, obtained from our k-dependent study.

The effects of Antimycin A, shown in Figure 6B, are almost identical to those of Nigericin. Similar to Nigericin, cells treated with Antimycin A do not produce mitochondrial ATP. Both types of treated cells show a decreased rate of long-range filament motion, suggesting that this motion is due to the action of

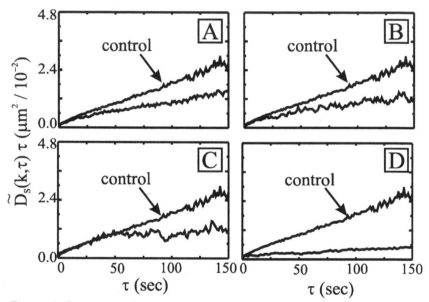

Figure 6. Comparison between FICS measurements on normal cells and cells treated with drugs that (A) uncouple respiration, (B) inhibit respiration, (C) depolymerize cytoskeletal actin, and (D) induce apoptosis. (Reproduced from reference 9. Copyright 2000 Biophysical Society.)

ATP-driven cytoskeletal filaments. The short-time short-range motion, however, is independent of metabolic activity.

Figure 6C shows results for cells treated with Latrunculin A that depolymerizes actin filaments, a major component of the cytoskeleton. For short-

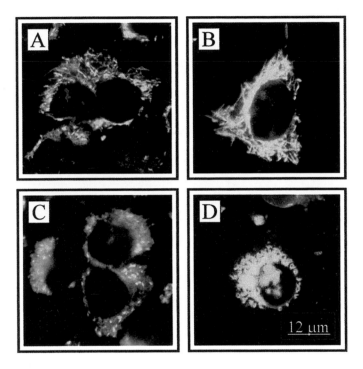

Plate 1. Effects of metabolism altering agents on JC-1 labeled cells. (A) Normal physiological conditions. (B) Uncoupled respiration. (C) Inhibition of respiration. (D) Initiation of apoptosis. (Reproduced from reference 11. Copyright 2000 Biophysical Society.)

times, $\tilde{D}_S(k,\tau)\tau$ is indistinguishable from control measurements, while for $\tau >$ 50 sec, the long-range motion is completely turned off. This suggests that the short time motion is a consequence of the mechanical properties of the membrane. The effects of apoptosis induced by Staurosporine are shown in Figure 6D. In this case, short- and long-time motions are dramatically reduced, consistent with the expected behavior of a swelled membrane. The absence of motion at long-times is consistent with the fact that ATP synthesis is shut down early in the apoptotic process.

Comparison to Other Methods

In comparison to microscopy, FICS has both important advantages and limitations. In essence, direct imaging experiments carry out many single-particle measurements in parallel. FICS experiments probe the time-course of collective fluctuations from an N-body system. While real-space trajectories contain all of the dynamical information that characterizes the system, this information must be statistically averaged to construct physically meaningful distribution functions. FICS data provide a direct route to the same relevant two-point distributions yielded by microscopy. Variations of FICS in which spatial information is simultaneously determined at more than one wave number at a time should provide the necessary information to compute higher-order spatial and temporal distributions. Such information may easily identify the signatures of non-uniform dynamics, without the necessity of measuring the full spatial distribution at once.

Acknowledgements

This work was supported by grants from the National Science Foundation (CHE-9876334 and CHE-9808049), the M. J. Murdock Charitable Trust (No. 98181), and the American Chemical Society Petroleum Research Foundation (No. 34285-G7).

References

1. Chaikin, P. M.; Lubensky, T. C. In *Principles of Condensed Matter Physics;* Cambridge University Press: Cambridge, 1995; pp 353-354.
2. Berne, B. J.; Pecora, R. In *Dynamic Light Scattering;* Krieger: Malabar, 1976.
3. Boon, J. P.; Yip, S. In *Molecular Hydrodynamics;* Dover: New York, 1991.

70

4. Knowles, M. K.; Grassman, T. J.; Marcus, A. H. *Phys. Rev. Lett.* **2000,** *85,* 2837-2840.
5. Grassman, T. J.; Knowles, M. K.; Marcus, A. H. *Phys. Rev. E* **2000,** *60,* 5725 — 5736.
6. Fleming, G. In *Chemical Applications of Ultrafast Spectroscopy;* Oxford University Press: New York, 1986.
7. Marcus, A. H.; Schofield, J.; Rice, S. A. *Phys. Rev. E* **1999,** 60, 5725 – 5736
8. Crocker, J. C.; Grier, D. G. *J. Colloid Interface Sci.* **1996,** 179, 298.
9. Margineantu, D.; Capaldi, R. A.; Marcus, A. H. *Biophys. J.* **2000,** 79, 1833-1843.

Chapter 6

Transient Structures of Solids and Liquids by Means of Time-Resolved X-ray Diffraction and X-ray Spectroscopy

D. A. Oulianov, I. V. Tomov, and P. M. Rentzepis[*]

Department of Chemistry, University of California, Irvine, CA 92697–2025

We describe a time-resolved x-ray diffraction and EXAFS experimental system, suitable for the study of ultafast processes in liquids and solids, with picosecond time resolution. The system uses a laser pulse to excite the sample and a delayed ultrashort hard x-ray pulse, generated in an x-ray diode driven by the same laser, to probe the structure of intermediate species produced by the excitation. A few time-resolved experiments are described including the picosecond time-resolved x-ray diffraction of laser heated metal and semiconductor crystals, and time-resolved EXAFS of carbon tetrabromide in ethanol solution before and after photoinduced dissociation.

Introduction

Currently, ultrafast time-resolved optical spectroscopy remains the most popular technique for the study of ultrafast processes in physical, chemical and biological systems. It is based on the 1968 picosecond pump/probe technique (1): the process is initiated by a pump pulse, and then the changes in the sample are probed by a delayed probe pulse. In ultrafast time-resolved optical spectroscopy, the probe pulse is optical (it is either an ultrashort monochromatic laser pulse or a non-coherent pulse with a wide spectrum, e.g. supercontinuum laser pulse). The use of this technique provides the spectra and kinetics of intermediate species produced during the photoinduced process and by performing the appropriate quantum mechanical molecular calculations the structure of the intermediate states may be calculated. These calculations, however, are rather complicated and usually limited to just a handful of simple molecules. Nevertheless, in many cases the knowledge of the transient structures of the molecules involved in the core of a chemical or biological process is very important, if not mandatory, for understanding thoroughly the process occurred.

Ever since the discovery of the x-rays, x-ray diffraction has been the most extensively used method to study the structure of materials (2). The x-ray diffraction from a single molecular crystal provides the most accurate lattice parameters and the molecular structure information. This technique, however, requires very high order of sample periodicity, and therefore in most cases, cannot be used for accurate structure determination of amorphous solids and liquids. One popular method that can be used in these cases is extended x-ray absorption fine structure (EXAFS) spectroscopy (3). This method measures the x-ray absorption spectrum in the vicinity of the absorption edge of a specific atom. Analyzing the low amplitude oscillations in the x-ray absorption spectrum in the range of 40- 1000 eV energies higher than the absorption edge, one can calculate the structure of the first few coordination layers around the atom. There are several other methods not discussed in this paper, which have also been successfully applied for structure determination of both solids and liquids. They include the electron and neutron diffraction.

In the past, only static structures were determined by x-ray diffraction and EXAFS spectroscopy. In recent years, the advances in ultrashort pulsed X-ray sources have opened the possibility to extend these methods to the ultrafast time-resolved domain. Several ultrafast time-resolved x-ray diffraction experiments with nano- and picosecond time resolution have been reported recently (4-10) (excellent reviews of the most important results could be found in Ref. 11), however, to the best of our knowledge, no ultrashort time-resolved EXAFS experiment has ever been performed previously.

Several types of ultrafast x-ray sources have been used for time-resolved x-ray diffraction. They are laser generated plasma, laser driving photodiodes and pulsed synchrotron sources. The synchrotron sources are the most powerful sources, however they are not easily accessible and have pump/probe synchronization problems. To generate the ultrashort x-ray pulses we have used a laser driving diode. Because the generated x-ray beam is divergent the x-ray photon flux at the position of the sample is quite low. Therefore, the accumulation of many pulses is required for a reasonably good quality x-ray diffraction pattern or an EXAFS spectrum. In the x-ray diffraction experiments presented here, the typical exposure time was of the order of 1 hour when a 300 Hz repetition rate x-ray source was used. For the EXAFS data the average experimental collection time was about 100 hours. We expect to improve the experimental time required for future experiments by using x-ray monolithic polycapillary focusing optics, which we have just received for testing from X-ray Optical Systems Inc. (XOS). A similar system used in conjunction with a micro x-ray source was reported by XOS to result in an intensity gain of 4400 times at 8 keV and 2400 times at 17.4 keV (12).

In this report we present our ultrafast time-resolved x-ray diffraction and EXAFS experimental systems with picosecond time resolution. First we will present some of our time-resolved x-ray diffraction studies on laser induced heat and strain propagation in metal and semiconductor crystals with nanosecond and picosecond time resolution (5, 6). In the second part of this report we will present our data on structure determination of liquids, specifically on the EXAFS spectroscopy of carbon tetrabromide/ethanol solution before and after photoinduced dissociation (13).

Time-Resolved X-ray Diffraction

In this section we present an application of ultrafast time resolved x-ray diffraction for the study of lattice behavior during pulsed laser illumination, by means of time-resolved Bragg-profile measurements. When energy from a laser pulse interacts with a material, it generates a non-uniform transient temperature distribution, carrier concentration and other effects, which alter the lattice structure of the crystal. The deformed crystal lattice will change the angle of diffraction for a monochromatic x-ray beam by $\Delta\theta = -(\Delta d/d)\cdot\tan\theta_B$, where d is the spacing of the diffracting planes, Δd is the change of the spacing due to an outside influence, and θ_B is the Bragg angle. Thus, for a divergent incoming x-ray beam, the diffracted signal from the inhomogeneous crystal will consist of signals scattered over a range of angles which are related to the depth distribution of the strain. Using a CCD detector, we were able to detect the distribution of the diffracted x-ray beam along the Bragg angle coordinate with high resolution and consequently measure the lattice deformation accurately.

This suggests that the transient crystal structure changes, induced by low energy short laser pulses, can be measured directly.

The experimental system, which we used for time-resolved x-ray diffraction experiments, is shown in Figure 1. It consists of the laser system, which produces nanosecond or picosecond UV pulses, the laser driving x-ray diode and the detection system. The laser system has two functions: it excites the sample and drives the x-ray diode, which generates the x-ray pulses used to probe the changes in the sample. Detail description of the experimental system can be found elsewhere (14). For nanosecond x-ray diffraction experiments we employed the ArF excimer laser, which produces 193 nm laser pulses with duration of 12 ns (FWHM) at 300 Hz repetition rate. For picosecond diffraction the laser system consists of the ArF laser used as an amplifier seeded by 193 nm picosecond pulse generated by the dye laser system (see Ref. 14 for details). This system produces 193 nm laser pulses with duration of 1.8 ps (FWHM) and energy up to 0.5 mJ/pulse at 300 Hz repetition rate. The x-ray diode consists of two flat electrodes: copper anode and aluminum photocathode. The characteristic Cu Kα radiation ($\lambda = 1.54$ Å) is used for these measurements. The x-ray probe pulse duration was 12 ns and 8 ps for the nanosecond and picosecond systems respectively. In both experiments, the x-ray pulses generated by the x-ray diode, after passing through two parallel slits separated by 180 mm, were directed to the sample crystal at the Bragg angle for Cu Kα lines. The sample was mounted on a four-axis Eulerian cradle in order to be accurately oriented. The scattered x-ray radiation was recorded by a liquid nitrogen cooled 2Kx2K CCD camera (15 μm pixel). The CCD camera is made specifically for direct x-ray imaging, and the geometric resolution of this experimental system is about 15 μrad. The advantage of the large-area CCD detector is that it allows for the simultaneous recording of the reflected x-ray radiation from different parts of the sample. Thus, a signal from the laser interaction spot and a reference signal from not excited part of the sample are recorded at the same time.

First we have performed the nanosecond time-resolved x-ray diffraction experiment in order to study the deformation of the structure in GaAs (111) crystal after nanosecond-pulsed UV laser irradiation. The GaAs crystal absorbs strongly 193 nm radiation, and because only 10 nm of the surface is penetrated by the UV photons, the bulk of the crystal is heated by diffusion. The energy of the heating laser pulse on the crystal surface was about 3 mJ corresponding to an energy density of 30 mJ/cm^2. This is several times smaller than the melting threshold of GaAs, which was reported to be 225 mJ/cm^2 for nanosecond pulses at 193 nm (15). The size of the UV spot, on the crystal, was much larger than the x-ray penetration depth in the crystal, therefore one-dimensional distribution of the temperature and stress in the probed bulk of the crystal was assumed. Only very high quality GaAs(111) crystals were used in these experiment. Several crystals of varying sizes were cut from 0.5 mm thick, 50 mm diameter GaAs wafer. Figure 2 shows the experimental Kα1 and Kα2 rocking curve (the

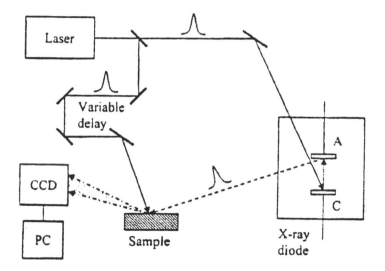

Figure 1. Experimental system for pump/probe time resolved x-ray studies. A, anode; C, cathode; (——) laser pulses; (----) x-ray pulses; (··········) electron pulses; (··—··-) diffracted x-ray pulses.

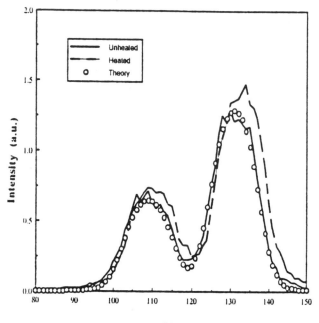

Figure 2. Experimental rocking curve for a cold and hot GaAs crystal. The delay time is 10 ns. The points are the calculated rocking curve for the cold crystal. Both Cu Kα1 (stronger) and Cu Kα2 lines are recorded. 1 pixel = 50 μrad.

intensity profile of the diffracted spot) for a cold and hot (measured at 10 ns after excitation) GaAs crystal. The shift and increase in intensity of the excited crystal rocking curve can be clearly seen. Figure 3 represents the measured and calculated integrated reflectivity for delays up to 50 ns after excitation. The increase in the integrated intensity of the rocking curve can be explained by the increase in the acceptance angle of the excited crystal due to small lattice deformations. These data show directly a histogram of the evolution of the transient structure of the crystal and its eventual return to the original lattice spacing. We have calculated the rocking curves for all delays used in this experiment (the dynamical equations for slightly deformed crystal were used, see Ref. 5 for details). By fitting the calculated and experimental rocking curves the lattice deformation profile in the crystal over a 100 ns time window was obtained (Figure 4).

We have also performed similar measurements for a Au(111) crystal using our picosecond time-resolved x-ray diffraction system. The gold crystal was 150 nm thick, grown on a 100 μm thick mica crystal. Electron diffraction patterns showed a well ordered Au(111) crystal over several mm parallel to the surface. Thus we assume that there is a mosaic structure along the surface, but along the thickness it is practically a single crystal.

Heating solid materials with picosecond laser pulses, when neither melting nor vaporization were induced, has been studied theoretically in detail (16). We have calculated the heat distribution in the gold crystal using the optical, thermal and other properties of gold and mica, which are relevant for our experiment. From these data we find that for a 1.8 ps laser pulse, the diffusion length is L_d = 22 nm. This length represents the crystal depth heated during the pulse illumination. From the above calculations we find that 0.1 mJ of absorbed energy in a spot size of S = 0.1 cm^2 will increase the temperature of the volume SL_d of gold by about 190°C. The heat from this volume spreads within picoseconds to the rest of the 150 nm thick crystal. According to the heat diffusion theory it takes about 90 ps for heat equilibrium to be established inside the SL crystal volume. The heat dissipation from this volume may take three directions: along the gold film, to the air or through the mica substrate.

The above estimates show that for the first 100 ps after the heat inducing laser pulse, the UV irradiated area of the crystal is in a non-equilibrium transient stage. Inside the crystal there is a thermal strain associated with a heated surface layer. Since the heated spot is much larger than the crystal thickness, it is reasonable to assume a one-dimensional strain distribution.

A temperature gradient generated in the crystal lattice will alter the x-ray diffracted pattern. In our experiments the x-ray pulse is probing the entire thickness of the crystal. The absorption loss, for the diffracted x-rays, traveled the longest distance through the gold crystal is about 30%. Therefore, the recorded diffracted pattern is an integration, over the probed crystal volume, for

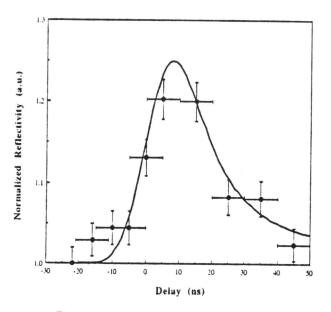

Figure 3. Calculated integrated reflectivity of laser pulse heated GaAs(111) crystal as a function of time The points are the experimental results.

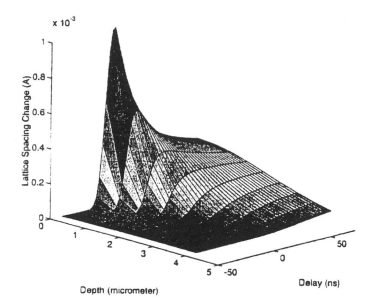

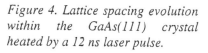

Figure 4. Lattice spacing evolution within the GaAs(111) crystal heated by a 12 ns laser pulse.

the time of the x-ray pulse duration. We used the theory of x-ray scattering from a one dimensionally strained crystal to calculate the diffracted x-ray intensity for a given temperature distribution. Figure 5 shows the experimentally measured and calculated rocking curves of the Au(111) crystal for several time delays before and after laser excitation. In Figure 6, the shift of the peak of the rocking curve as a function of delay time is shown. The transition through a thermally nonuniform crystal lattice, in the first 50 ps, is clearly seen. In this transition time, the width of the rocking curve is also slightly larger than the one at equilibrium. The spread of the experimental points is partially due to the shot to shot fluctuations in the UV pulse energy which results from the jitter in the triggering of the excimer amplifier. After about 100 ps and up to 500 ps, which was the longest delay used in this experiment, no change in the shift was observed. These measurements show conclusively that our experimental system is capable of 10 ps time resolution and can easily detect transient lattice structure deformations caused by temperature changes of about $20^{\circ}C$.

Time-Resolved EXAFS

EXAFS spectroscopy is based on the analysis of low amplitude oscillations in the region of 40-1000 eV higher than the absorption edge of a specific atom in the x-ray absorption spectrum of the material. After absorption of an x-ray photon with the energy E exceeding the edge energy E_0, the atom most probably emits a photoelectron with a kinetic energy equal to $E_k=E-E_0$. The photoelectron wave propagates from the absorber atom and scatters from the orbital electrons of the surrounding atoms. The scattered electron waves travel back to the absorber atom and interfere with the outgoing electron wave resulting in modulations in the absorption cross section of the absorber atom. Analyzing the EXAFS spectrum one can calculate the structure information of a few shells surrounding the absorber atom. This information includes the average distances between the absorber and the scatterers, the average square deviations from these distances (which result from the thermal oscillations and disorder of the system), the coordination number for each of the shells and in some cases even the high order anharmonic force cumulants.

The raw EXAFS data are represented by the dependence of an x-ray mass absorption coefficient times the length of the sample μx ($\mu x = \ln(I_0/I)$, where I_0 is the intensity before the sample and I is the intensity after the sample) on the energy of the x-rays. In order to get a pure EXAFS spectrum the continuum background, which represents the absorption spectrum of a single separated atom, is removed from the raw EXAFS spectrum. The result is normalized and the x-ray energy is converted into the absolute value of a photoelectron wavevector k ($h^2k^2/2m = E-E_0$, where h is a Planc constant and m is an electron

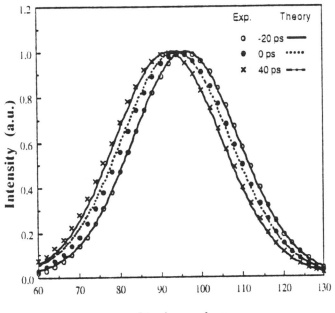

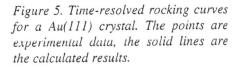

Figure 5. Time-resolved rocking curves for a Au(111) crystal. The points are experimental data, the solid lines are the calculated results.

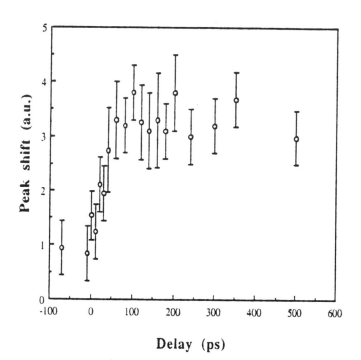

Figure 6. The shift of the rocking
curve as a function of delay time.

mass). The calculated pure EXAFS spectrum $\chi(k)$ is processed further by application of Fourier transform, which converts it to a radial distribution function $F(r)$. The structural parameters of the surrounding atoms can be calculated by fitting the theoretically calculated radial distribution function to the experimental one. Very often, especially in the cases when several shells are detected, it is convenient to backtransform the $F(r)$ into k space with a window, which filters the oscillations associated only with one of the shells, and perform the fitting for each of the shells separately.

Figure 7 shows the schematic diagram of the ultrafast time-resolved EXAFS system. The laser system and the x-ray diode are the same as the ones used for the nanosecond time-resolved x-ray diffraction experiments described in the previous section, with the only difference that a copper anode was replaced by a tungsten one. The x-ray continuum generated by our laser pumped pulsed x ray diode has the same energy spectrum as that of a conventional tungsten x-ray tube. The divergent nanosecond x-ray continuum pulse was formed by a 50 μm slit situated 6 mm away from the anode. After passing through a 2 mm cell containing liquid sample, the x-ray pulse was reflected from a 20 cm diameter Si(111) crystal oriented for (422) reflection. The center of the crystal is situated 20 cm away from the slit. An x-ray CCD detector (1242x1152 pixels) is situated 40 cm away from the crystal and is capable of recording, simultaneously, about 500 eV of the x-ray spectrum. In the course of the experiment one half of the x-ray beam passes through the sample while the other half propagates through air only and is used as a reference. This arrangement makes possible the simultaneous recording of both the sample and reference EXAFS signals. Tungsten Lγ lines were used for both calibration and to determine the 5 eV resolution of the x-ray spectrometer in the energy region of the Br edge (13.5 keV). By selecting the crystal and decreasing the slit even a higher energy resolution is possible, however at the expense of the x-ray flux.

In order to test our system we have recorded the bromine K-edge EXAFS spectra of 0.4 M aqueous solution of $ZnBr_2$. Figure 8 shows the experimental raw EXAFS data. A standard automated data reduction procedure (code AUTOBK, see Ref. 17) was used to extract the pure EXAFS spectrum. The EXAFS spectrum was also calculated theoretically using the FEFF8 code (18). It was shown that the dominant signal for the first few hundreds eV of the EXAFS spectrum of $ZnBr_2$ in aqueous solution is derived from the nearest oxygen atoms, and the scattering from Zn^{++} ions at this energy range can be neglected (19). Therefore for our calculations we used only Br-O scattering paths. The fitting was performed using the FEFFIT code (20) in the backtransformed k-space using $r = 1.3$-4.0 Å filtering window. Figure 9 shows the experimental data and the theoretically calculated EXAFS spectrum in the backtransformed k-space. From the fitting we have calculated the average distance between Br⁻ ions and the

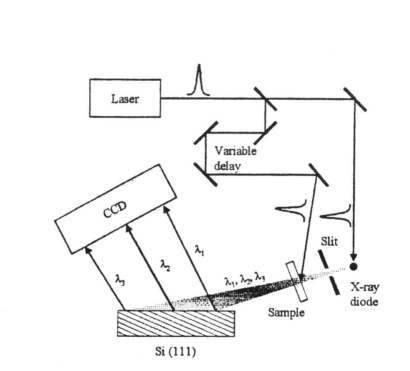

Figure 7. Schematic diagram of the time-resolved EXAFS experimental system.

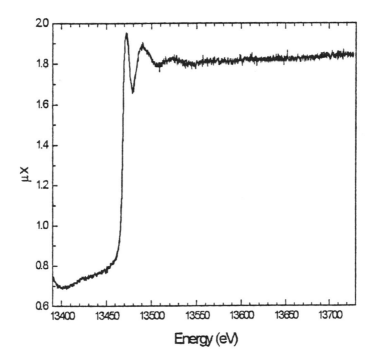

Figure 8. Br K-edge EXAFS spectrum of ZnBr₂/water solution.

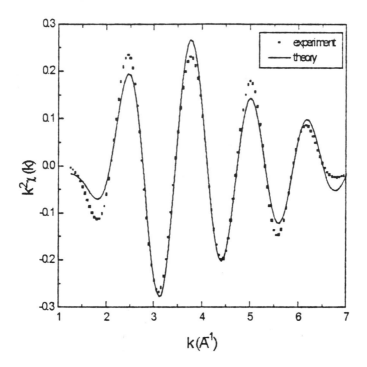

Figure 9. Br K-edge EXAFS spectrum of ZnBr₂ after the first shell filtering. The points are the experimental results, the solid line is theoretical calculations.

oxygen atoms in the first solvation shell: $r = 3.28\pm0.02$ Å, which agrees with the previously reported value within the experimental error (19).

After the successful EXAFS experiment with $ZnBr_2$/water solution we also performed EXAFS experiments on 0.05 g/ml carbon tetrabromide ethanol solution before and after UV irradiation. Figure 10 shows the radial functions of the irradiated and not irradiated samples. It is clear from the picture that the structure surrounding the Br atom before and after irradiation is very different. We have performed 2 shell fitting for the non-irradiated sample in the backtransformed k-space (see Figure 11) using theoretical calculations for molecular CBr_4. The distances were found to be $r = 1.98\pm0.01$ Å (for C-Br distance) and $r = 3.17\pm0.01$ Å (for Br-Br distance). These results differ from the previously reported distances (21) by 0.04 Å and 0.02 Å for C-Br and Br-Br distances respectively. This can be probably explained by the fact that in Ref. 21 samples in gas phase were used, while we have studied the samples in solution. We have also tried to fit the UV irradiated CBr_4 data using several simple models with a single type of atom in the first shell. The results were unsuccessful. This suggests that a more complicated model, with several types of atoms in the first shell and probably with consideration of multiple scattering paths, is required.

We would like to emphasize that the EXAFS data reported here is only preliminary. We have studied starting and final products of the photodissociation reaction of carbon tetrabromide in ethanol solution without measuring the structures and kinetics of intermediate transient products. Work on these experiments is in progress (13). Charge-transfer complexes of bromine atoms with haloalkanes and alkanes produced by laser-flash photolysis and pulse radiolysis in various organic solvents have been studied for many years (22, 23). The laser photolysis process of CBr_4 takes several possible pathways before the final stable products are obtained. The time frame of these intermediate reactions are found to vary widely. In alcohol solution the CBr_4 dissociation may take the following roots:

$$CBr_4 + h\nu \rightarrow CBr_3\cdot + Br\cdot \quad ,$$
$$CBr_4 + Br\cdot \rightarrow CBr_4\cdot \ Br \ ,$$
$$Br\cdot + CH_3CH_2OH \rightarrow CH_3\dot{C}HOH\cdot + HBr \ ,$$
$$CH_3\dot{C}HOH\cdot + CBr_4 \rightarrow CH_3CHBrOH + CBr_3\cdot \quad ,$$
$$CBr_3\cdot + CH_3CH_2OH \rightarrow CHBr_3 + CH_3\dot{C}HOH\cdot \quad .$$

The dissociation of CBr_4 occurs within picoseconds or less however the $CBr_3\cdot$ $CBr_4\cdot$ Br and $CH_3CHOH\cdot$ will be long lived (longer than 1 ns) and easier measured by nanosecond time-resolved EXAFS if their concentration is high.

The above results show that the change in the structure of carbon tetrabromide induced by photodissociation can be clearly seen with our

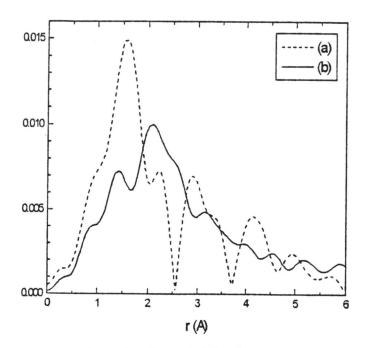

Figure 10. The radial distribution function of CBr₄/ethanol solution: (a) before UV irradiation, (b) after UV irradiation.

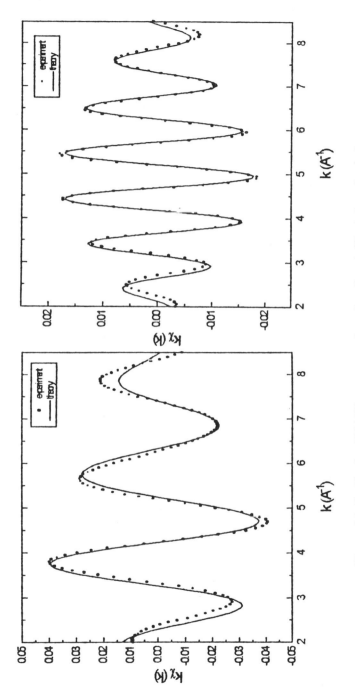

Figure 11. Br K-edge EXAFS spectrum of not irradiated CBr₄/ethanol solution after (a) the first shell filtering; (b) the second shell filtering. The points are the experimental results, the solid line is theoretical calculations.

90

dispersive time-resolved EXAFS system. This data also demonstrated that our experimental system can be used for ultrafast time-resolved EXAFS experiments. However to determine accurate structure information with the existing experimental system exposure times of over 100 hours may be required. We have strong evidence that we will improve the system by increasing the x-ray flux on the sample by more than a factor of 100, therefore the data collection time required for accurate structure determination of intermediate species will be significantly decreased.

Conclusion

We have developed an ultrafast time-resolved x-ray diffraction and EXAFS experimental system with time resolution of up to several picoseconds, which can be used for the study of a time-resolved structure change of excited and intermediate states in solids and liquids. This system has already been applied to study the heat and stress dynamics in photoexcited metal and semiconductor crystals by means of time resolved x-ray diffraction. We have also studied the bromine K-edge EXAFS spectra before and after photoinduced dissociation of carbon tetrabromide in solution by time-resolved EXAFS spectroscopy.

Acknowledgement

This work was supported in part by NSF grant # CHE-9501388 and ARO grant # DAAD19-00-1-0427.

References

1. Rentzepis, P. M. *Chem. Phys. Lett.* **1968**, 2, 117-120.
2. *X-ray Diffraction*; Warren, B. E.; Dover Publishers, Inc.: New York, 1990.
3. *X-ray Absorption*; Koningsberger, D. C.; Prins, R., Eds.; John Wily & Sons: New York, 1988.
4. Tomov, I. V.; Oulianov, D. A.; Chen, P.; Rentzepis, P. M. *J. Phys. Chem.* B **1999**, 103, 7081-7091, and references therein.
5. Chen, P.; Tomov, I. V.; Rentzepis, P. M. *J. Chem. Phys.* **1996**, 104, 10001-10007.
6. Chen, P.; Tomov, I. V.; Rentzepis, P. M. *J. Phys. Chem. A* **1999**, 103, 2359-2363.
7. Srajer, V.; Teng, T.; Ursby, T.; Pradervand, C.; Ren, Z.; Adachi, S.; Schildkamp, W.; Bourgeous, D.; Wuff, M.; Moffat, K. *Sience* **1996**, 274, 1726-1729.

8. Perman, B.; Srajer, V.; Ren, Z.; Teng, T.; Pradervand, C.; Ursby, T.; Bourgeous, D.; Schotte, F.; Wuff, M.; Kort, R.; K. Hellingwerf, K.; Moffat, K. *Sience* **1998**, 279, 1946-1950.

9. Chin, A. H.; Schoenlein, R. W.; Glover, T. E.; Balling, P.; Leemans, W. P.; Shank, C. V. *Phys. Rev. Lett.* **1999**, 83, 336-339.

10. Cavalleri, A.; Siders, C. W.; Brown, F. L. H.; Leitner, D. M.; Toth, C.; Squier, J. A.; Barty, C. P. J.; Wilson, K. R. *Phys. Rev. Lett.* **2000**, 85, 586-589.

11. *Time Resolved X-ray and Electron Diffraction;* Helliwell, J. R.; Rentzepis, P.M., Eds.; Oxford University Press: Oxford, 1997.

12. Gao, N.; Ponamarev, I. Yu.; Xiao, Q. F.; Gibson, W. M.; Carpenter, D.A. *Appl. Phys. Lett.* **1997**, 71, 3441-3443.

13. Oulianov, D. A.; Tomov, I. V.; Lin, S. H.; Rentzepis, P. M. *J. Chin. Chem. Soc.* **2001**, 48 (in publication).

14. Tomov, I. V.; Chen, P.; Lin, S. H.; Rentzepis, P. M. in *Time Resolved X-ray and Electron Diffraction;* Helliwell, J. R.; Rentzepis, P. M., Eds.; Oxford University Press: Oxford, 1997; Chapter 1.

15. Solis, J.; Afonso, C. N.; Piqueras, J. *J. Appl. Phys.* **1992**, 71, 1032-1034.

16. Bechtel, J. H. *J. Appl. Phys.* **1975**, 46, 1585-1593.

17. Newville, M.; Livins, P.; Yacoby, Y.; Rehr, J. J.; Stern, E. A. *Phys. Rev. B* **1993**, 47, 14126-14131.

18. Ankudinov, A. L.; Ravel, B.; Rehr, J. J.; Conradson, S. D. *Phys. Rev. B* **1998**, 58, 7565-7576.

19. Lagarde, P.; Fontaine, A.; Raoux, D.; Sadoc, A.; Migliardo, P. *J. Chem. Phys.* **1980**, 72, 3061-3069.

20. Newville, M.; Ravel, B.; Haskel, D.; Rehr, J. J.; Stern, E. A.; Yacoby, Y. *Physica* B **1995**, 208&209, 154-156.

21. Yokoyama, T.; Yonamoto, Y.; Ohta, T. *J. Phys. Soc. Japan* **1996**, 65, 3901-3908.

22. Lian, C.; Shoute, T.; Neta, P. *J. Phys. Chem.* **1990**, 94, 2447-2453.

23. Alfassi, Z. B.; Huie, R. E.; Mittal, J. P.; Neta, P.; Shoute, L. C. T. *J. Phys. Chem.* **1993**, 97, 9120-9123.

Chapter 7

Cooperative Dynamics in Polymer Liquids

Marina G. Guenza

Department of Chemistry, University of Oregon, Eugene, OR 97403

The dynamics of molecular liquids is driven by the complex interplay between intermolecular and intramolecular forces. While theoretical approaches to the dynamics of simple liquids use a detailed description of the intermolecular interactions, the dynamics of polymer fluids is assumed to be mainly intramolecular in nature, with the intermolecular forces traditionally approximated by a mean-field description. This theoretical approach reproduces the experimental data in the long-time regime, while it is in disagreement with the short-time polymer dynamics as measured in experiments and computer simulations. To overcome this problem we derived from the first-principle Liouville equation, by means of the Mori-Zwanzig projection operator technique, a new Generalized Langevin Equation that explicitly contains intramolecular and intermolecular contributions to the single chain dynamics. We argue that intermolecular contributions are responsible for the onset of anomalous dynamical behavior. The single polymer dynamics calculated with our approach is in good agreement with the observed dynamics over the full range of time scales investigated.

The dynamics of homogeneous dense fluids display complex behavior due to the presence of dynamical heterogeneities generated by interconverting regions of highly cooperative motions with varying mobility (*1,2,3*). This so-called "anomalous dynamics" is revealed by the behavior of the particle center-of-mass mean-square displacement, $\Delta R^2(t) \propto t^\nu$ with $\nu \neq 1$. For time intervals where $\nu < 1$, particle self-displacements appear to be caged by the presence of surrounding particles resulting in hindered, subdiffusive behavior. For long time intervals, a so-called "hopping transport mechanism" is observed leading to apparent directed motion, with $\nu > 1$.

Anomalous dynamics has been observed in both experiments and computer simulations on low molecular weight polymer fluids in the short-time regime (t < τ_{Rouse}, where τ_{Rouse} is the longest molecular relaxation time) (4,5,6). In these studies, the degree of polymerization $N \ll N_e$ where N_e is the entanglement degree of polymerization. Initially the single molecule center-of-mass mean-square displacement scales quadratically with time ($\nu = 2$). During intermediate time intervals, this ballistic dynamics crosses over to the long-time (t ≥ τ_{Rouse}) Fickian diffusion ($\nu = 1$) through an intermediate subdiffusive regime ($\nu = 0.83$). Only at long time intervals (t ≥ τ_{Rouse}) is the system free to diffuse resulting in the recovery of Fickian behavior.

The appearance of a subdiffusive regime is in disagreement with the conventional theoretical approach to the dynamics of low molecular weight polymer fluids, i.e., the Rouse model (7). In the Rouse model, the single chain dynamics is driven by intramolecular entropic restoring forces and segmental friction. The surrounding fluid is described as a continuum, serving as a thermal reservoir. As a consequence, each molecule is free to diffuse through the liquid which is structurally and dynamically uniform, and $\Delta R^2(t)$ is predicted to scale as a linear function of time ($\nu = 1$) for all times. The Rouse model agrees with the data only at long time (t ≥ τ_{Rouse}), where it also correctly describes the scaling of the bulk viscosity and the single-chain diffusion coefficient with N ($D_{Rouse} = k_B T/(\zeta N)$), where k_B is the Boltzmann constant, T is the temperature, and ζ is the monomer friction coefficient).

The Rouse model also fails to predict the dynamics of entangled polymer fluids. For times that are short in comparison to the entanglement relaxation time, τ_e, the dynamics of the polymer chain is not affected by the presence of surrounding chains and should follow the Rouse equation (7). Nevertheless, computer simulations of entangled polymer fluids show that, in this regime, the dynamics is subdiffusive with $\nu = 0.75 - 0.83$ (8,9,10,11). This short-time subdiffusive behavior appears in semidilute polymer solutions only above the polymer overlap concentration, c^* (8,9,10,11). This observation supports the hypothesis that the short-time anomalous dynamics is a consequence of intermolecular interactions. Anomalous subdiffusive dynamics was also observed in simulations of small gas molecules diffusing through polymer matrices (12).

Inconsistencies between the Rouse equation and experimental data are not surprising. First, the Rouse equation models local intermolecular forces through a simple mean-field approach. In reality, each molecule in a fluid experiences an effective intermolecular potential due to its local environment called the potential of mean force (*13*). The range of this potential is determined by the spatial extent of the correlation hole (*14*) in the molecular site-to-site pair correlation function g(r). g(r) is the distribution of distances separating two monomers belonging to two separate chains, averaged over all possible monomer positions and all possible chain conformations. The length scale of the correlation hole is of order R_g in polymer fluids (*15*). It is reasonable that two polymer chains whose center-of-mass separation is less than the range of R_g will affect each others dynamics. In polymer fluids, each individual molecule occupies a volume $V \propto R_g^3$, where $R_g^2 = N \, l^2/6$ is the polymer mean square radius of gyration, and l is the statistical segment length. Inside this volume are contained an average of $n \propto \sqrt{N}$ chains, that interact with each other through the potential of mean force (*13*). These polymer-polymer interactions are expected to significantly modify the dynamics on a length scale shorter than or equal to the range of the potential ($r \leq R_g$), and on a time scale shorter than or equal to the time necessary for the molecule to escape outside of the range of the potential ($t \leq \tau_{Rouse}$). For this reason, the Rouse equation is unlikely to give an accurate description of the short time ($t \leq \tau_{Rouse}$) polymer dynamics.

A second important drawback of the Rouse equation is that it neglects local polymer stiffness. The structure of a single chain is described as a collection of freely jointed segments of effective length, l. The presence of local semiflexibility is not relevant for the dynamics at large scales, so that the global dynamics of a semiflexible chain is equivalent to the global dynamics of a flexible chain of renormalized statistical segments, when they have the same radius of gyration. On the contrary, the local dynamics is substantially modified when the description of the chain statistics on a local scale is changed.

With the goal of overcoming the inconsistencies of the Rouse description we developed a new microscopic approach (*16*) to the dynamics of polymer fluids explicitly including intermolecular contributions. The theory is then extended to treat the dynamics of polymer with local stiffness. From an analysis of the normal modes of motion we see that the intermolecular forces mainly affect the dynamics on the global scale, while of the local scale the dynamics is unperturbed. As a consequence the single chain dynamics presents anomalous center-of-mass diffusion, while the segment relaxation is mainly affected by the presence of local semiflexibility. Both these results are in agreement with the data from experiments and computer simulations.

Theoretical background

The starting point in developing a description of the heterogeneous dynamics of a structurally homogeneous molecular fluid is to select a set of slow relevant dynamical variables (*13*). We chose as relevant variables the coordinates of the molecules undergoing slow cooperative dynamics (*1,2,3*). The fast dynamical variables are approximated by a mean-field description and perturb the single molecule dynamics through the random forces and the effective monomer friction coefficient. The well-defined difference in the time scales observed experimentally for the dynamics of the slow and of the fast variables, makes this approximation well justified for this type of system.

The choice of the slow relevant variables allows the mathematical construction of a projection operator. By projecting the dynamics of the whole fluid described by the Liouville equation onto the phase-space defined by the coordinate of the n molecules undergoing slow cooperative dynamics (*13*), we derived (*16*) a new Generalized Langevin Equation (Correlated Dynamics Generalized Langevin Equation). The final CDGLE is obtained by factorizing the many-body potential into intramolecular and pair-wise additive intermolecular site-site interactions. This CDGLE contains explicitly the contributions from the intermolecular potential of mean-force and can be extended to treat finite size polymer chains with local intramolecular stiffness.

The final equation of motion for the dynamics of an effective segment, a, in the molecule, i, is driven by the intramolecular potential $(-\beta^{-1} \ln[\psi(\{r\}^{(i)}(t))]$ with $\{r\}^{(i)}(t)= r_1^{(i)}(t), r_2^{(i)}(t), .. r_a^{(i)}(t), .. r_N^{(i)}(t)$ the ensemble of cartesian coordinates defining the segment positions inside the molecule i), and the intermolecular potential of mean-force, $-\beta^{-1}\ln[g(\{r\}^{(k)}(t), \{r\}^{(l)}(t))]$. The random interactions with the surrounding "fast" molecules are expressed by the projected random force (*13*), $F_a^{Q(i)}(t)$. The segment position $r_a^{(i)}(t)$ follows the equation of motion:

$$\varsigma \frac{dr_a^{(i)}(t)}{dt} = \frac{1}{\beta} \frac{\partial}{\partial r_a^{(i)}(t)} \ln\left[\prod_{j=1}^{n} \Psi\left(r^{(j)}(t)\right) \prod_{k<j}^{n} g\left(r^{(j)}(t), r^{(k)}(t)\right) \right] + F_a^{Q(i)}(t), \quad (1)$$

with $\beta = (k_B T)^{-1}$. The new equation of motion applies to the overdamped dynamics of a fluid of interacting molecules of any kind, including polymers. It recovers the traditional single-molecule Generalized Langevin equation in the limit n = 1, with the fluid described as a structureless continuum, g(r) = 1. In the case n = 2 and N = 1, it recovers a model for binary interacting particles extensively studied in the past (*17*). Thus, our equation of motion applies, as it must, in the two opposite and well-known limits of strongly dominant intramolecular or intermolecular interactions.

For polymer fluids, the intramolecular and intermolecular memory functions enter the effective monomer friction coefficient giving a correction proportional to N/N_e. For unentangled polymer dynamics this contribution becomes negligible (16) and it is discarded in this study.

In the special case of polymer molecules we take advantage of the Gaussian shape of the intramolecular distribution function and derive an average approximate analytical form for the intermolecular force acting between two tagged chains, i and k. Considering the n interacting chains inside the mean-force potential

$$\frac{1}{\beta} \frac{\partial}{\partial \mathbf{r}_a^{(i)}(t)} \sum_{k=1}^{n-1} \ln\left[g(\mathbf{r}^{(i)}(t), \mathbf{r}^{(k)}(t))\right] \approx \frac{n-1}{n} G(t)\left(\mathbf{r}_a^{(i)}(t) - \mathbf{R}_{c.m.}^{(k)}(t)\right)$$

with $G(t) = G(0) \exp[-3\, R^2(t) / (4\, R_g^2)]$ the intermolecular time-dependent interaction strength. $G(0) = -\rho\sqrt{(3\pi)}/\beta\ [3/2g(d) + \lambda\ \ln(\pi^2\rho^2 l^4 R_g^2/18)\ 9/\rho\pi l^2)]$ depends on the physical (density, ρ, and temperature) and chemical parameters of the system. When $\lambda = 1$ the potential is repulsive in agreement with the formal derivation of the pair distribution function for polymer fluids (15), while for $\lambda = -1$ the potential has a long attractive tail, in agreement with many empirical representations of intermolecular interactions, including the Lennard-Jones potential (13). By increasing the polymer molecular weight, $G(0)$ increases. An increase of the density of the system also enhances $G(0)$.

The time dependence of $G(t)$ follows from its explicit dependence on $R(t)$, and $G(t)$ becomes vanishingly small when the distance between the two molecules exceeds several R_g. The anomalous dynamics is confined to time scales $t \leq \tau_{Rouse}$, which is the time necessary for the system to travel outside the range of the potential, $R^2(t) \geq R_g^2$. Introducing the intramolecular distribution function and $G(t)$ in eq 1, we obtain a set of non-linear Rouse equations with explicit intermolecular contributions,

$$\zeta \frac{d\mathbf{r}_a^{(i)}(t)}{dt} = k_s \frac{\partial^2 \mathbf{r}_a^{(i)}(t)}{\partial a^2} - \frac{n-1}{n} G(t)\mathbf{r}_a^{(i)}(t) + \frac{n-1}{n} G(t)\mathbf{R}_{c.m.}^{(k)}(t) + \mathbf{F}_a^{Q(i)}(t), \quad (2)$$

where k_s is the entropic intramolecular spring constant. Equation 2 is solved by transformation to normal mode coordinates. With the purpose of comparing the theory with the available data, we translate the equation into a matrix form that allows the calculation of the dynamical properties for a melt of finite size semiflexible polymers (Optimized Intermolecular Rouse Approximation).

Model calculations and comparison with experimental data

Unentangled polymer melts center-of-mass diffusion

In general the increase of $G(0)$ leads to the emergence of complex dynamical behavior. In Figure 1 we demonstrate this effect on the mean-square displacement of the single-chain center-of-mass, $\Delta R^2(t)$. $\Delta R^2(t)$ is calculated numerically through a self-consistent procedure to take into account the change of the interactions with increasing intermolecular distance. In a few limiting cases $\Delta R^2(t)$ can be reduced to a simple analytical form. For weak intermolecular interactions, $G(0) \rightarrow 0$, and $\Delta R^2(t)$ recovers Rouse diffusion. For strongly interacting polymers the dynamics presents anomalous diffusion in a regime confined to time $\tau_1 \le t \le \tau_{Rouse}$ with $\tau_1 = \zeta/(2\,G(0))$. In general $\tau_1 \ne \tau_0$, where τ_0 is the free collision time (*13*).

If we assume that the infinitesimal translation of the center-of-mass is linear in time, we obtain an analytical equation for the center-of-mass dynamics. For an attractive long-range potential the equation shows two well-defined types of dynamics. For long, slowly relaxing polymer melts, $9\,D\,t < 2\,R_g^2$, and $G(0) > 0$ an anomalous slowing down of the dynamics takes place for $\tau_1 \le t \le \tau_{Rouse}$ (Figure 1). Our physical explanation of this anomalous behavior (*15*) is that the molecule is temporarily trapped due to the intermolecular potential of the neighboring molecules, and anomalous slowing down of the dynamics takes place, $\Delta R^2(t) \propto t^\nu$ with $\nu < 1$. At longer time, $t \approx \tau_{Rouse}$, the molecules undergo many-chain correlated dynamics, $\Delta R^2(t) = 6\,D_{m.c.}t$ with $D_{m.c.}=(\beta\,n\,N\zeta)^{-1}$. When the thermal energy of the system is sufficiently high in comparison to $G(0)$, $\beta G(0) \rightarrow 0$, at time $t=\tau_d$ the molecule recovers free diffusion, re-establishing the effective ergodicity of the system.

For $G(0) >> 0$ and $\tau_d >> \tau_{Rouse}$, the system is not able to recover single-chain dynamics during the time scale investigated. The dynamically correlated domains persist in the system, which appears to behave non-ergodicly. Thus an increase of $G(0)$ leads to an apparent ergodic to non-ergodic transition.

The theory predicts that the long-time ($t \approx \tau_{Rouse}$) correlated many-chain dynamics takes place only if $9\,D\,t < 2\,R_g^2$ and $G(0) > 0$. For short, highly mobile molecular fluids, $9\,D\,t > 2\,R_g^2$, and the dynamics is diffusive. It follows that anomalous "caged" diffusion and correlated many-chain dynamics are predicted to be a characteristic features of undercooled systems ($D \rightarrow 0$) (*18*), and/or entangled polymer fluids ($N >> N_e$) (*7*). Eventually at long enough times $9Dt > 2R_g^2$ and single-chain dynamics is also recovered in slowly moving, high molecular weight polymer fluids.

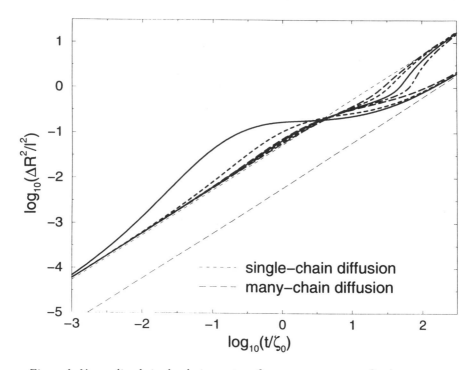

Figure 1. Normalized single chain center of mass mean-square displacement vs normalized time, at decreasing strength of the attractive interaction, G'(0). From the top curve to the bottom curve (close to the single chain free diffusion) G'(0)=G(0), G'(0)=0.2G(0), G'(0)=0.1G(0), G'(0)=0.09G(0), G'(0)=0.08G(0), G'(0)=0.067G(0), G'(0)=0.05G(0). G(0) is the intermolecular interaction at initial conditions: $T=1\ k_B^{-1}$, $N=100$, $n=\sqrt{N}$, $\rho=3l^3$, and $R^2(0)=R^2_g$.

If the long-range potential is repulsive, $G(0) < 0$, and the many-chain correlated dynamics does not take place. The dynamics crosses over from the anomalous subdiffusive regime to Fickian single-chain diffusion. The subdiffusive regime becomes less relevant for short, highly mobile polymer chains.

To test the quantitative features of these predictions we compare the theory with the simulated data of $\Delta R^2(t)$ and monomer mean-square displacement for a melt of unentangled polyethylene chains (4,5), n-$C_{100}H_{202}$, (entanglement degree of polymerization $N_e = 136$) (Figure 2). These simulations have been shown to agree with spin-echo experiments on polyethylene melts at $T = 509$ K, and with the Rouse long-time dynamics when the polymer is described as a flexible chain of $N = 14$ statistical segments (4,5). Each segment has an effective bond length l $= 1.047$ nm that reproduces the experimental radius of gyration, $R_g = 1.6$ nm. From

$$n(r) = N^{-1} \int_0^r dr'\, \rho g(r')$$

we estimate the number of interacting polymer chains inside the range of the potential ($r \approx R_g$) to be $n \approx 12$. This value is obtained by assuming an average intermolecular center-of-mass distance at time zero $R^2(0) \approx R_g^2$, as calculated from

$$R^2(0) \propto \int_0^{R_g} n(r) r^2 dr$$

From the long-time diffusion coefficient, corresponding to the regime where the polymer follows single-chain Rouse dynamics, we derive the value of the monomer effective friction coefficient, $\zeta = 0.2 \times 10^{-9}$ dyne s/cm, which agrees with the experimental value (19). We use ζ together with the experimental T, ρ, N, and l, as an input to the theory. We keep as the only fitting parameters the number of polymer chains effectively correlated, n', and the strength of the intermolecular potential, $G'(0)$. Quantitative agreement between eq 2 with a repulsive potential, and the simulated center-of-mass mean-square displacements is obtained in the whole range of data for $n' \approx 1.6/n$ (Figure 2), where n is the "theoretical" value previously defined, and with $G'(0) \approx 5G(0)$ for $g(d) = 4$, or $G'(0) \approx 10G(0)$ for $g(d) = 0$. The slight discrepancy between the predicted and the calculated values can be ascribed to the approximations implicit in our analytical solution of eq 1.

The qualitative and quantitative agreement between our approach and the data supports the hypothesis that in polymer fluids the appearance of anomalous dynamics is related to the presence of local intermolecular interactions.

Unentangled polymer melts monomer diffusion

The monomer mean-square displacement, $\Delta r^2(t)$, as calculated in computer simulations of unentangled polymer melts, appears to follow closely Rouse dynamics (4,5). This behavior suggests that in a polymer melt the dynamics is less affected on a local scale than on a global scale by the presence of intermolecular forces, in agreement with experimental findings (20,21). This latter prediction is consistent with the usual "static" picture of a homopolymer melt, where the polymer structure is unperturbed as a consequence of the similar and compensating intramolecular and intermolecular excluded-volume interactions (14). Unperturbed local dynamics was also found (6) in simulations of the local segment reorientation time-correlation function, $P_2(t)$. An unperturbed Rouse local dynamics was found in the simulated short-time monomer dynamics of entangled polymer melts (9,11).

The simulations of unentangled PE melt dynamics (4,5), however, present a slightly different scaling exponent of the monomer mean-square displacement in the intermediate region where the Rouse model predicts $\Delta r^2(t) \propto t^{0.5}$. This discrepancy is related to the presence of local polymer semiflexibility. To investigate the importance of the local stiffness on polyethylene dynamics we use a freely rotating chain model for $N = 100$ segments with an effective stiffness parameter $g = 0.74$ [$g = <l_i \cdot l_{i+1}>/l^2$, with $l_i = (r_{i+1} - r_i)$] which reproduces the experimental characteristic ratio for a finite polyethylene chain. Our approach, extended to describe semiflexible polymer chains of finite size, predicts a slightly slower than Rouse short-time monomer dynamics that better agrees with the simulation data (Figure 2). The center-of-mass dynamics, instead, is not affected by the local polymer stiffness since in our comparison the overall polymer dimension is fixed to the experimental value.

Unentangled undercooled polymer melt dynamics

As a second test of our approach we performed a comparison with computer simulation data of center-of-mass diffusion for unentagled undercooled polymer melts with $N=10$. The input parameters are from the computer simulation (18), while the fitting parameters are the number of dynamically correlated molecules, n, and the strength of the intermolecular interaction, G(0). Figure 3 shows that the fitting procedure is in quantitative agreement with the data, with reasonable

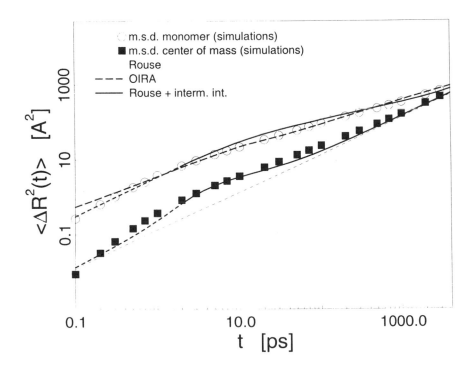

Figure 2. Monomer and center of mass mean-square displacements as a function of time. Best fit of the molecular dynamics simulation data with the Rouse equation, and with the Rouse equation including repulsive intermolecular interactions for a flexible chain of N=14 renormalized statistical segments, l=1.047 nm. Best fit with a semiflexible chain of N=100 segments, l=0.153 nm and g=0.74, following Rouse dynamics with intermolecular interactions (OIRA).

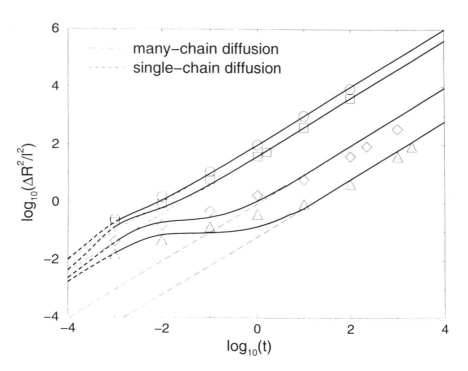

Figure 3. Best fit of the simulation data of center of mass mean-square displacement as a function of time, for a unentangled polymer melt at decreasing temperature: $T=1\ k_B^{-1}$ (circle); $T=0.4\ k_B^{-1}$ (square); $T=0.23\ k_B^{-1}$ (diamond); $T=0.19\ k_B^{-1}$ (triangle). The effective intermolecular potential is assumed to be repulsive at $T=1\ k_B^{-1}$ and $T=0.4\ k_B^{-1}$, and attractive at $T=0.23\ k_B^{-1}$ and $T=0.19\ k_B^{-1}$.

values of the fitting parameters, n and G(0). At high temperature the dynamics is driven by repulsive interactions, where $n = 13$ and $G'(0) = 3G(0)$ for $k_B T = 1$, and $n = 15$ with $G'(0) = 10G(0)$ for $k_B T = 0.4$ for $g(d) = 0$. These values of n are consistent with eq 3, and our previous calculation of the center-of-mass dynamics for a PE chain of $N = 14$ statistical segments freely jointed together. Close to the melt state the dynamics follows subdiffusive behavior that crosses over to free single-chain diffusion. No cooperative many-chain dynamics takes place.

When the temperature is further decreased, the dynamics follows a qualitatively different mechanism. A strong decrease is found in the long time diffusion coefficient, which follows the empirical Vogel-Fulcher law. From our calculations this regime emerges as a diffusive process involving the cooperative dynamics of many chains, whose number, n, dramatically increases with decreasing temperature. A good fit of the data is found with the bare monomer friction coefficient kept constant from the high T regime, while the best fitting parameters are $n = 25$ with $G'(0) = 0.02G(0)$ at $k_B T = 0.23$, and $n = 300$ with $G'(0) = G(0)$ at $k_B T = 0.19$. The intermolecular potential in the low temperature regime is assumed to be attractive. An attractive effective intermolecular potential is possible when the time of relaxation of the dynamical heterogeneities exceeds the time of observation. Our fitting parameters indicate a dramatic increase of the range of the dynamical correlation with decreasing temperature in agreement with the experimental findings (2). This effect corresponds to the observed longer-range oscillations in the mean-force potential approaching the glass transition, and suggests that the freezing of the dynamics could be related to a rapid increase in the number of polymer chains involved in the cooperative many-chain dynamics with decreasing temperature.

Conclusion

We presented a new theoretical approach to the dynamics of structurally homogeneous, but dynamically heterogeneous molecular fluids. We have shown that the presence of the intermolecular mean-force potential, which is discarded in conventional approaches, induces anomalous single-molecule diffusion much like that observed in experiments and computer simulations. Anomalous dynamics is confined to the spatial scale corresponding to the range of the effective potential, which in polymer fluids it is of the order of the polymer radius of gyration.

While the conventional Rouse equation cannot reproduce the experimental anomalous diffusion of unentangled polymer dynamics, our extended Rouse equation with repulsive intermolecular forces is in quantitative agreement with data in the complete range of time- and length-scales investigated. The theory also qualitatively explains the anomalous dynamics in undercooled unentangled polymer fluids with a repulsive potential at high temperature and an attractive potential close to T_g.

Our approach predicts that the presence of intermolecular forces mainly affects the melt dynamics on the global spatial scale, while the local dynamics is unperturbed. It is only when the intermolecular interactions become long ranged and the dynamics becomes highly cooperative that the behavior is also perturbed on the local spatial scale. This occurs when the polymer fluid approaches its glass transition, or at the crossover from unentagled to entangled dynamics.

Above the glass transition, the monomer mean-square displacement follows the unperturbed dynamics in agreement with the Rouse model. The slight discrepancies of the computational data of monomer diffusion with the Rouse scaling exponent for a melt of polyethylene chains can be traced to the presence of the local polymer stiffness. This new approach, extended to treat finite semiflexible polymer chains, is found to reproduce quantitatively the monomer diffusion data.

Acknowledgement

We are grateful to W.Paul for sharing his computer simulation data. Acknowledgment is made to the donors of The Petroleum Research Fund, administrated by the ACS, for partial support of this research. We also acknowledge the support of the National Science Foundation under the grant DMR-9971687.

Reference

1) Donati, C.; Glotzer, S.C.; Poole, P.H. *Phys. Rev. Lett.* **1999**, *82*, 5064-5067.
2) Marcus, A.H.; Schofield, J.; Rice, S.A. *Phys. Rev. E* **1999**, *60*, 5725-5736.
3) Weeks, E.R.; Crocker, J.C.; Levitt, A.C.; Schofield, A.; Weitz, D.A. *Science* **2000**, *287*, 627-631.
4) Paul, W.; Smith, G.D.; Yoon, D.Y. *Macromol.* **1997**, *30*, 7772-7780.
5) Paul, W.; Smith, G.D.; Yoon, D.Y.; Farago, B.; Rathgeber, S.; Zirkel, A. Willner, L.; Richter, D. *Phys.Rev.Lett.* **1998**, *80*, 2346-2349.
6) Kopf, A.; Dunweg, B.; Paul, W. *J. Chem. Phys.* **1997**, *107*, 6945-6955.
7) Doi, M.; Edwards, S.F. *The Theory of Polymer Dynamics;* International Series of Monographs on Physics 73; Publisher: Oxford Science Publications, Oxford, GB, 1994.

8) Kremer, K.; Grest, G.S. *J. Chem. Phys.* **1990**, *92*, 5057-5086.

9) Paul, W.; Binder, K.; Heermann, D.W.; Kremer, K. *J. Chem. Phys.* **1991**, *95*, 7726-7740.

10) Kolinski, A.; Skolnick, J.; Yaris, R. *J. Chem. Phys.* **1987**, *86*, 1567-1585.

11) Smith, S.W.; Hall, C.K.; Freeman, B.D. *J. Chem. Phys.* **1996**, *104*, 5616-5637.

12) Muller-Plathe, F.; Rogers, S.C.; van Gusteren, W.F. *Chem. Phys. Lett.* **1992**, *199*, 237-243.

13) Hansen, J.-P.; McDonald, I.R. *Theory of Simple Liquids;* 2nd ed.; Publisher: Academic Press, London, GB, 1991.

14) De Gennes, P.-G. *Scaling Concepts in Polymer Physics;* 4th ed.; Publisher: Cornell University Press, Ithaca, NY, 1993.

15) Schweizer, K.S.; Curro, J.G. *Adv. Chem. Phys.* **1997**, *98*, 1-142.

16) Guenza, M. *J. Chem. Phys.* **1999**, *110*, 7574-7588.

17) Vesely, F.J.; Posch, H.A. *Mol. Phys.* **1988**, 64, 97-109.

18) Okun, K.; Wolfgardt, M.; Bashnagel, J.; Binder, K. *Macromol.* **1997**, 30, 3075-3085.

19) Pearson, D.S.; Fetters, L.J.; Graessley, W.W.; Ver Strate, G.; von Meerwall, E. *Macromol.* **1994**, *27*, 711-719.

20) Richter, D.; Willner, L.; Zirkel, A.; Frago, B.; Fetters, L.J.; Huang, J.S. *Phys. Rev. Lett.* **1993**, *71*, 4158-4161.

21) Richter, D.; Willner, L.; Zirkel, A.; Frago, B.; Fetters, L.J.; Huang, J.S. *Macromol.* **1994**, *27*, 7437-7446.

Photochemistry

Chapter 8

Electron Photodetachment in Solution

Jeremiah A. Kloepfer, Victor H. Vilchiz, Victor A. Lenchenkov,
and Stephen E. Bradforth

Department of Chemistry, SSC 702, University of Southern California,
Los Angeles, CA 90089

The mechanism for photoejection of electrons from simple inorganic anions in aqueous solution is being explored using ultrafast UV pump - visible/IR probe spectroscopy in close connection with quantum simulations.   The pathway for detachment in the prototype aqueous iodide system, excited with a single photon into the quasi-bound charge-transfer-to-solvent (CTTS) electronic state, is probed in detail. Our experiments observe, for the first time, the timescale of ejection of the electron into the solvent from the lowest CTTS state, and the subsequent relaxation of the solvent to accommodate and solvate the electron. The ejection process is compared with resonant detachment of the molecular anion $[Fe(CN)_6]^{4-}$. These low energy ejection pathways are contrasted with multi-photon ionization of solutes, or the solvent itself, into the conduction band.

Introduction

Photoionization and photodetachment, terms that describe the process of ejecting an electron from a neutral and negative ion respectively, can be rather simply described when they occur in vacuum. On exceeding a certain threshold when tuning the incident photon energy, electrons are ejected. If additional energy is delivered in the absorbed photon, this ends up in excess kinetic energy of the outgoing photoelectron. Except at threshold, the outgoing electron can be treated as leaving the atomic or molecular parent suddenly, and the probability of the electron being close to its parent at any subsequent time is vanishingly small. For ejection of electrons in the liquid phase, even at this crude level of description, the mechanism of electron ejection is rather different and much less clear. The general picture of electron ejection by ionizing radiation in water starts with a band structure picture of the solvent. If the ionizing radiation has sufficient energy to promote an electron from the solvent, or from a dissolved solute impurity, the electron is promoted into the conduction band. However, a band picture for an inhomogeneous liquid is relatively poor and it is recognized that the electron in the conduction band is not extensively delocalized. In fact, the electron is rapidly localized and becomes trapped in the solvent at some distance from the ionization center. Once the surrounding solvent has a chance to relax its local structure, a solvated electron is formed. The latter particle is very long lived (μs-timescale) and essentially diffuses in the liquid like an atomic ion. The distance between the initially trapped electron and its ionized parent is known as the thermalization length, and in multiphoton ionization of water, the average distance is ~ 10 - 15 Å; this value depends on the total energy delivered by the incident photons (1). Multiphoton detachment of anions ejecting electrons via this type of pathway has also been studied (2,3).

However, it has long been known that there are pathways to production of a solvated electron that involve photoexcitation at much lower energy. The best known is the photodetachment of aqueous halide anions through prominent charge-transfer-to-solvent (CTTS) bands in the ultraviolet. Several aqueous aromatic neutrals exhibit similar behavior. The prototype iodide system's lowest band peaks at 225 nm (5.6 eV), some 1.6 eV below the vertical threshold to reach atomic iodine in water and a vacuum electron. The precise mechanism of this threshold detachment process is the subject of this paper. Helping in the goal of attaining a detailed microscopic description of this process are two recent developments: quantum state resolved simulations of the electron detachment dynamics for aqueous I^- from Rossky's group employing hybrid quantum-classical non-adiabatic molecular dynamics (4) and evidence for analogous excited state spectroscopy (5-7) and ejection dynamics (8) in small hydrated iodide clusters in the gas phase. We hope to show that rather a detailed description of this threshold ejection mechanism is now in fact emerging.

For iodide ions in bulk water at room temperature, the first solvent shell contains on average 8 - 9 waters with the water $O - I^-$ distance averaging 3.6 Å (9). Although no fixed hydration shell should be inferred, on average the water molecules are oriented with one hydrogen pointed toward the anion. When the

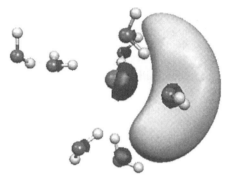

Figure 1. *Snapshot of a molecular dynamics simulation of aqueous iodide employing 864 H_2O molecules at room temperature. Only the closest water molecules are shown. At this instantaneous geometry, an ab initio calculation is performed to find the lowest triplet excited state – the water molecules in the first solvent shell (shown) are fully quantum-mechanically treated and the remaining 856 waters are treated as point charges. The black and gray lobes shown are isodensity surfaces of opposite phase for the molecular orbital with the promoted electron; the electron fills space opened up by fluctuation of the water network. We note that the excited state formed when iodide is vertically excited by the laser pulse is a singlet state, however the HOMO is expected to be similar. The CTTS wavefunction is thus defined by the instantaneous asymmetry of the environment; the electron will undergo complete detachment with the subsequent solvent response to the excitation to become a solvated electron*

5p valence electron in I^- is excited, there exist excited states where the orbital of the promoted electron is bound by the pre-existing solvation structure of the water environment. We note that bare iodide has no bound excited states in the gas phase. With just a few water molecules, the excited state is bound (by a few tenths of an eV) by the dipole moment of the asymmetric water cluster (*5-7*), however in the bulk it is the existing extended polarization of the medium at the instant of excitation that binds the promoted electron (by ~ 1.6 eV for I^-) (*10,11*). Figure 1 shows the shape of a typical orbital populated by CTTS resonant excitation for an instantaneous configuration of the water medium. As shown in Rossky's simulations, once the water nuclei have time to respond to the change in electron density on the iodide solute, the CTTS state collapses (*4*). The CTTS state, although bound with respect to a vacuum electron and the vertical solvent configuration, is not stable with respect to nuclear rearrangement and detachment of the electron to become trapped elsewhere in the solvent. Therefore, the electron transfer into solvent is controlled by the timescale of the solvent motions.

Our experiments are designed to test this picture for threshold photodetachment and answer some simple questions, for example, how long does it take the electron to separate from its parent and where does it trap. We have explored the resonant photodetachment of aqueous iodide via CTTS in detail and compared it to two other ejection processes, that of two photon ionization of water itself at 9.7 eV (sufficient energy to reach the conduction band) and resonant photodetachment of $[Fe(CN)_6]^{4-}$. The 2-photon ionization of water has been studied recently by a number of groups, however our experiment has the highest time resolution to date and provides a one-to-one comparison to experiments on anion detachment. The $[Fe(CN)_6]^{4-}$ system has long been employed as a photolytic source of hydrated electrons due to its high quantum yield at relatively long wavelengths. However, the mechanism of electron production and the interplay of overlapping valence electronic transitions is not well understood.

Results and Discussion

I^- photodetachment

Experiments in our lab employ a 50 fs tunable UV pump, broadband probe spectrometer to follow the photoejection dynamics of aqueous solutions (*12,13*). Because of the very rapid timescales of water motions, it is important to use the shortest pulses possible to resolve the detachment dynamics – a challenge at the deep UV wavelengths necessary for one photon detachment. A pump pulse at 255 nm excites I^- or $[Fe(CN)_6]^{4-}$, in aqueous solution, into their lowest CTTS state, or at high intensities drives two photon ionization of the pure solvent (*13-15*). A subsequent probe pulse detects the appearance of a solvated electron in the solution by its transient absorption. The solvated electron absorbs across the entire visible spectrum and into the near IR (*16*), and conveniently is the only absorbing species in this spectral range. Thus, we probe right across the electron spectrum using a white light continuum. For example, Figure 2 shows the raw experimental signal for I^- detachment at three probe wavelengths. There are three phases of evolution from which we can extract information about the ejection dynamics. The initial rise (~ 200 fs, significantly longer than our instrument response) indicates the delayed appearance of the solvated electron.

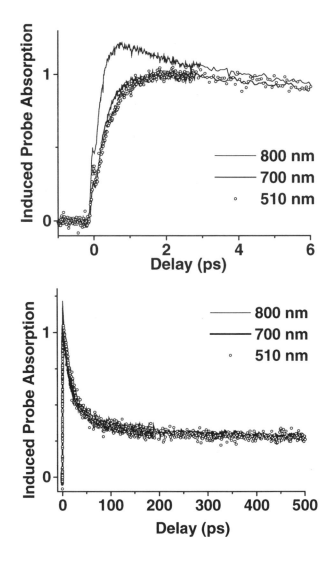

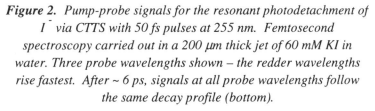

Figure 2. *Pump-probe signals for the resonant photodetachment of I⁻ via CTTS with 50 fs pulses at 255 nm. Femtosecond spectroscopy carried out in a 200 μm thick jet of 60 mM KI in water. Three probe wavelengths shown – the redder wavelengths rise fastest. After ~ 6 ps, signals at all probe wavelengths follow the same decay profile (bottom).*

Initially, the signal varies for the different probe wavelengths, indicating a spectral evolution phase. After ~ 6 ps, all probes exhibit the same evolution – the long decay is due to geminate recombination of the product electron with its iodine parent (*13*). Interestingly, pump-probe anisotropy experiments at all probe wavelengths find no memory in the generated electrons for the initial polarization direction of the detachment pulse.

By using all probe wavelength data we can determine a two-dimensional map (see Fig. 3) of the earliest phase of the ejection dynamics. The map clearly shows a continuous spectral shifting of the electron spectrum and no evidence for a two state transition from a precursor electron state, at least within our probe spectral window. Furthermore, the map provides a fingerprint of the environment the electron finds itself in after ejection (*14,17*). We have analyzed these maps with a global fitting approach to separate the appearance time of the electron from the timescale for relaxation due to solvent rearrangement as evidenced by the continuous spectral shift. However, even qualitative inspection of the experimental maps indicates a contrasting picture of ejection and solvation from the three distinct ionization pathways studied. For iodide, a ground state electron is formed on a 200 fs timescale and the electron spectrum shifts with a timescale of 850 fs. For water ionization, the ground state electron is also formed in ~200 fs. However, the rearrangement of the water molecules surrounding the electron that leads to spectral shifting can be seen (Fig. 3) to be considerably faster. The characteristic time constant from the global fit is indeed faster by about a factor of two, and the extent of the spectral shift is larger (0.56 eV vs. 0.36 eV for I^-) (*14*). The solvation time for the photoelectron formed by water ionization is similar to the longitudinal dielectric relaxation timescale for water. The slower solvation timescale observed for the *detached* electron is suggestive of a different local environment surrounding this photoelectron.

It is clear from Figure 2 and 3 that by 3 – 6 ps, the ground state spectrum of the electron is no longer shifting and the pump-probe signal therefore reflects pure population evolution. For delays < 1 ns, the electron population is decaying by geminate recombination with the detached/ionized parent species which was also created in the ionization event. The transient absorption signal may then be directly equated with the geminate pair survival probability function, and this used to extract the distance to which the electron was initially ejected. In particular, the longer the initial thermalization length, the larger the fraction of electrons that escape recombination altogether and the slower the kinetics of those that recombine. Analytic solutions for the pair survival probability, which are distinctly non-exponential, are available for diffusion limited and partially diffusion limited reactions and parameterized in terms of the initial pair radial distribution (*18,19*). For bulk water ionization at 9.7 eV, a good fit for our experimental population decay is obtained assuming diffusion limited recombination of the photoelectron starting out at an average radius of 15 Å. The

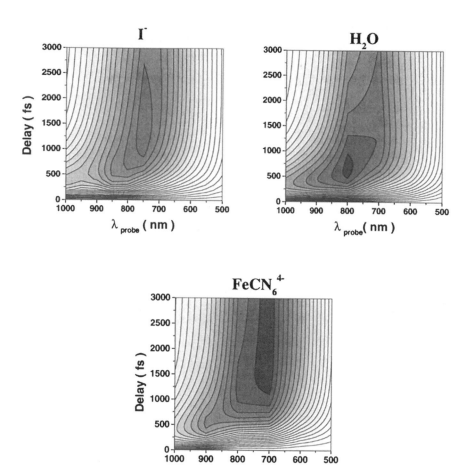

Figure 3. *Probe spectral map of nascent photoelectron provides a fingerprint of the appearance and relaxation dynamics of the ejected electron. Each contour plot shows the experimental signal intensity (darkest is highest transient absorption intensity) as a function of pump-probe delay and probe wavelength. These plots are constructed from up to 11 probe datasets such as those shown in Fig. 2. The three plots shown compare ejection for I^- and $[Fe(CN)_6]^{4-}$ via 1 photon CTTS detachment and H_2O via two-photon ionization. Bottom panel reproduced with permission from reference 15. Elsevier 2001.*

recombination partner is predominantly an OH radical which is believed to be rapidly formed at the ionization site after rapid proton transfer from H_2O^+ to a neighboring water. Further, the reaction takes place on-contact at a separation of 5.7 Å (the latter value is independently fixed from the bulk OH + e^- rate constant once the reaction is assumed diffusion limited) (13). This result is in excellent agreement with the recent literature. Applying a diffusive model, with an electron returning from "long" range to recombine, to the I^- photodetachment data, however, fails to yield a satisfactory fit.

Quantum/classical molecular dynamics simulations for halide detachment suggest a radically different picture (4,20,21) - that detachment of the electron from the CTTS state leads to a halogen/electron caged pair. Further, the simulations suggest that there is an attractive interaction between the halogen and electron in the cage that implies diffusive escape from the cage is activated and that the non-adiabatic return electron transfer (ET) reaction is fairly slow (21). In this limit, the survival probability of the electron is determined by competition between return ET within the caged pair to form ground state iodide and activated diffusive escape of the electron from the pair. A kinetic form with only three adjustable parameters was proposed by Staib and Borgis to describe the survival probability function (21). A fit of our time-resolved iodide data with this equation yields good results (Fig. 4), with 33 ps and 70 ps time constants fitted for non-adiabatic recombination and escape, respectively (13). It is apparent that for time scales longer than ~200 ps, the simple kinetic model levels off and deviates from the experimental data. This is reasonable: the competitive kinetics model omits the possibility that electrons that diffuse out of the pair may still return for secondary (successful) encounters. When such secondary recombination is taken into account, now using a numerical solution of the diffusion equation including an electron-iodine attractive interaction potential (21), an excellent fit (Figure 4) to the data is obtained over the complete time range measured (22). By systematically varying the viscosity with added cosolvent we have further verified this model – the slower the electron escapes from the contact pair due to reduced mobility, the higher the recombining fraction.

Our analysis of the photodetachment experiments of iodide leads to the following conclusions. The excitation of the CTTS state leads to substantial rearrangement of the solvent shell surrounding the I atom in the first 200 fs, "budding" the excess electron into a caged pair. This contrasts with the situation for multi-photon ionization where the electron is ejected through a spatially extended conduction band, rapidly trapping into a distant solvent site. Our conclusion is based on both the geminate recombination profile and the relaxation map of the detached electron, which shows significantly slower solvation for the ejected photoelectron, supporting the assignment of the presence of the iodine neutral atom in the immediate solvent cage. Time-

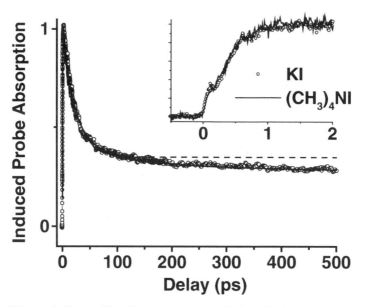

Figure 4. *Recombination of electron with detached iodine atom, assuming return electron transfer competes with escape from initially formed caged pair. Data (circles) recorded with 700 nm probe. Two models are shown: (dashed) competitive kinetics model and (solid) numerical solution of the diffusion equation from a initially formed contact pair having attractive potential well (depth 580 cm⁻¹). (Inset) early time electron signal evolution at 700 nm showing invariance of rising edge to nature of counter-ion. The transient on the rising edge is instrument limited and appears at all probe wavelengths (14).*

resolved scavenging data *(22)* from our lab further support this conclusion. This analysis points to a rather unexpected result. Despite the presumably excellent electronic overlap, the I:e⁻ pair is metastable, with a slow $(33 \text{ ps})^{-1}$ return electron transfer rate, k_n. Recent experiments on detachment of Cl⁻ and Na⁻ in Laubereau and Schwartz's labs respectively come to similar conclusions (Table I) *(23,24)*. Our best hypothesis is that the return ET is in the Marcus inverted regime and the rate varies depending on the energy gap for charge recombination in the different radical atom/electron pairs *(4)*. For each of these atomic electron acceptors, there can be no internal "promoting modes" to enhance the inverted regime electron transfer rate. Thus, the overall long-time yield for solvated electrons in monovalent anion CTTS-type detachment is determined by the return electron transfer rate, the caged pair potential energy well depth and the mutual diffusion coefficient of the pair in the solvent.

Table I. Average electron thermalization lengths after multi-photon ejection and threshold ejection via CTTS.

Target	Excitation	Thermalization length	Reference
H_2O	9.7 eV conduction band	$r_0 = 15$ Å	(13)
I^-	12 eV conduction band	$r_0 \sim 15$ Å	(2)
I^-	4.8 eV CTTS	pair $k_n = 33$ ps	(13)
Cl^-	8 eV CTTS	pair $k_n > 70$ ps	(23)
Na^- (THF)	1.5eV CTTS	pair $k_n = 1.5$ ps	(24)

We have completed a series of experiments addressing how changing the local environment influences detachment. This includes varying the counter ion, increasing the solution ionic strength or including hydrogen bond breakers (*e.g.*, sucrose) in the aqueous environment. We find that the initial appearance and relaxation of the detached electron shows surprisingly little sensitivity to these factors, all of which would be expected to influence the local solvent structure. For example, the inset to Figure 4 shows there is no effect on the signal rising edge when the salt counter ion is changed from K^+ to $N(CH_3)_4^+$.

$[Fe(CN)_6]^{4-}$ photodetachment

To examine the generality of the CTTS detachment mechanism, we have performed a detailed study of the $[Fe(CN)_6]^{4-}$ system (*15*). Aqueous hexacyanoferrate(II) yields solvated electrons on one-photon absorption out to wavelengths as long as 313 nm, but the quantum yield is strongly dependent on excitation wavelength (Figure 5) (*25*). In contrast to the situation for I^-, there are a number of overlapping valence electronic transitions in addition to an assigned CTTS transition throughout this spectral region (*26*). The repulsive Coulomb potential for an electron departing and returning to its parent, the intramolecular degrees of freedom, the change of symmetry in the parent anion as well as the mixed nature of the initially excited state are all expected to play a role in the dynamics.

Our pump-probe experiments reveal the following: the ejected electron relaxation captured via the probe spectral map (Fig. 3) indicate a characteristic solvation time scale for the ground state electron of ~570 fs and an accompanying blue shift of 0.53 eV. Interestingly, this is in close agreement with our result for electrons ejected from water (rather than electrons undergoing photodetachment from iodide) suggesting that the electron is trapped and undergoing solvation at a remote site from the photooxidized iron complex. The appearance time of the electron in the solvent (~ 310 fs) is further delayed compared to ejection from I⁻ or from water. In contrast to previous reports, geminate recombination of the electron with iron (III) is indeed observed despite the opposite charges on the recombination partners. We find that that the recombination fraction varies strongly with solution ionic strengths (Fig. 5(b)). This stands in contrast to our result for iodide, which shows no variation in the observed recombination up to high ionic strength. However, these results can be readily rationalized by considering the extent of counter-ion pairing in the equilibrium precursor in water. Iodide ions do not ion-pair even at very high ionic strengths, however, at only modest ionic strengths, the $[Fe(CN)_6]^{4-}$ ion is associated with one or more K^+ counter-ions (27). Thus by varying the solution ionic strength, we are tuning the magnitude of the Coulomb repulsion (or the effective charge on the detached parent species). At the highest ionic strengths studied (4 M), we may assume that the electron is effectively recombining with a *neutral* oxidized iron complex associated with three cations. Based on a full analysis of the time-resolved electron recombination as a function of ionic strength, employing numerical simulations that include the presence of a repulsive Coulomb potential and ion pairing, we can unravel the Coulomb effect and extract the average ejection distance. Unlike the one-photon CTTS detachment processes in Table 1, the electron appears to be ejected from $Fe(CN)_6^{4-}$ to relatively long range (~ 15 Å) and this, rather than the Coulomb repulsion effect on geminate recombination, mainly accounts for the high photolytic solvated electron quantum yield. Once again, this is consistent with the 570 fs solvation timescale recovered from the probe map.

Finally, to attempt to understand the origin of the strongly varying quantum yield with exciting wavelength (Fig. 5(a)), we photodetached the $[Fe(CN)_6]^{4-}$ system with one photon at three pump wavelengths across its absorption spectrum (15,28). We found identical dynamics for appearance, relaxation and recombination despite the pump energy varying by ~0.75 eV. This is rather surprising given the nature of the overlapping electronic valence transitions in this region and/or the energy available to the excess electron. We believe this will allow us to reassign the aqueous $[Fe(CN)_6]^{4-}$ absorption spectrum, with a broad CTTS absorption underlying the entire UV spectrum. The quantum yield then merely follows the fraction of molecules excited directly into the CTTS

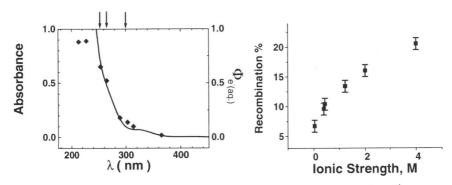

Figure 5. *(Left) The absorption spectrum of aqueous [Fe(CN)₆]⁴⁻* and the solvated electron absolute quantum yield (diamonds, ref. 25). The arrows indicate pump wavelengths for our experiments. **(Right)** The long time escape fraction of electrons after photodetachment of [Fe(CN)₆]⁴⁻ as a function of solution ionic strength. Ionic strengths is varied either by increasing concentration of K₄[Fe(CN)₆] alone, or by addition of spectator ions from KBr. The datapoints plotted are for 255 nm detachment, however, we find no variation with pump wavelength. Right panel reproduced with permission from reference 15. Elsevier 2001.*

band and the ejection dynamics need not necessarily require internal conversion between valence and CTTS states as has hitherto been prescribed (25,29).

Conclusion

We find that photodetachment of monovalent anions such as iodide proceed by electrons "budding" into a contact pair, akin to molecular photodissociation inside a cage, rather than by ejection into the bulk solvent. The dynamics of this initial process are determined by the timescale of rearrangement of the first hydration shell. Interestingly, there is still a significant yield of solvated electrons that escape the cage because of the anomalously slow return ET reaction with atomic iodine. The formation of a caged atom/electron pair is consistent with quantum/classical molecular dynamics simulations. For the first multiply charged anion we have studied, $[Fe(CN)_6]^{4-}$, the anion excited state from which electrons are ejected must be much more spatially extended as we observe comparable ejection distances to two-photon ionization of water. It turns out that the most important factor determining the solvated electron quantum yield for this system is not Coulomb repulsion limiting geminate recombination between the detached electron and its parent– rather it is the long initial ejection range. There is no variation with pump wavelength between 255 and 300 nm in the photodetached electron dynamics in the $[Fe(CN)_6]^{4-}$ system.

Barbara and coworkers have recently correlated the spatial extent of various states of the hydrated electron and precursor states to its static quenching capacity (30). They find that both the photoexcited $[Fe(CN)_6]^{4-}$ state and the initial state accessed by two photon excitation of water are both spatially extended (r ~ 15 Å), consistent with our findings from geminate recombination and solvation results for the electrons trapped after collapse of these intial states. Similar style static quenching experiments in our lab for iodide further confirm that the initial I^- CTTS wavefunction is much smaller .

Acknowledgments

The electronic structure calculations of bulk I^- CTTS states were performed in collaboration with Dr. Pavel Jungwirth. This research is supported by grants from the NSF and by the Donors of the Petroleum Research Fund administered by the American Chemical Society. S.E.B. is a recipient of the Camille and Henry Dreyfus Foundation New Faculty Award, a Cottrell Scholar of Research Corporation and a David and Lucile Packard Foundation Fellow in Science and Engineering.

References

1. Crowell, R. A.; Bartels, D. M. *J. Phys. Chem.* **1996**, *100*, 17940-17949.
2. Long, F. H.; Lu, H.; Shi, X.; Eisenthal, K. B. *Chem. Phys. Lett.* **1990**, *169*, 165.
3. Long, F. H.; Shi, X.; Lu, H.; Eisenthal, K. B. *J. Phys. Chem.* **1994**, *98*, 7252.
4. Sheu, W.-S.; Rossky, P. J. *J. Phys. Chem.* **1996**, *100*, 1295.
5. Serxner, D.; Dessent, C. E. H.; Johnson, M. A. *J. Chem. Phys.* **1996**, *105*, 7231.
6. Chen, H.-Y.; Sheu, W.-S. *J. Am. Chem. Soc.* **2000**, *122*, 7534-7542.
7. Majumdar, D.; Kim, J.; Kim, K. S. *J. Chem. Phys.* **2000**, *112*, 101-105.
8. Lehr, L.; Zanni, M. T.; Frischkorn, C.; Weinkauf, R.; Neumark, D. M. *Science* **1999**, *284*, 635-638.
9. Ohtaki, H.; Radnai, T. *Chem. Rev.* **1993**, *93*, 1157-1204.
10. Blandamer, M.; Fox, M. *Chem. Rev.* **1970**, *70*, 59.
11. Bradforth, S. E.; Jungwirth, P. *J. Phys. Chem. A.* **2001**, *in press*.

12. Kloepfer, J. A.; Vilchiz, V. H.; Lenchenkov, V. A.; Bradforth, S. E. *Chem. Phys. Lett.* **1998**, *298*, 120-128.
13. Kloepfer, J. A.; Vilchiz, V. H.; Lenchenkov, V. A.; Germaine, A. C.; Bradforth, S. E. *J. Chem. Phys* **2000**, *113*, 6288-6307.
14. Vilchiz, V. H.; Kloepfer, J. A.; Germaine, A. C.; Lenchenkov, V. A.; Bradforth, S. E. *J. Phys. Chem. A* **2001**, *A 105*, 1711-1723.
15. Lenchenkov, V. A.; Vilchiz, V. H.; Kloepfer, J. A.; Bradforth, S. E. *Chem. Phys. Lett.* **2001**, *342*, 277-286.
16. Jou, F.-Y.; Freeman, G. R. *J. Phys. Chem.* **1979**, *83*, 2383-2387.
17. Vilchiz, V. H.; A.Kloepfer, J.; Germaine, A. C.; Lenchenkov, V. A.; Bradforth, S. E. *Ultrafast thermalization dynamics of hot photoelectrons injected into water*; Elsaesser, T., Mukamel, S., Murnane, M. M. and Scherer, N. F., Ed.; Springer-Verlag: Berlin, 2000.
18. Tachiya, M. *Radiat. Phys. Chem.* **1983**, *21*, 167- 175.
19. Rice, S. A. *Diffusion-Limited Reactions*; Elsevier: Amsterdam, 1985; Vol. 25.
20. Sheu, W.-S.; Rossky, P. *J. Chem. Phys. Lett.* **1993**, *202*, 186.
21. Staib, A.; Borgis, D. *J. Chem. Phys.* **1996**, *104*, 9027-9039.
22. Kloepfer, J. A.; Vilchiz, V. H, Lenchenkov, V. A.; Chen, X.; Bradforth, S. E. *J. Chem. Phys.* **2001**, *unpublished*.
23. Assel, M.; Laenen, R.; Laubereau, A. *Chem. Phys. Lett.* **1998**, *289*, 267-274.
24. Barthel, E. R.; Martini, I. B.; Schwartz, B. J. *J. Chem. Phys.* **2000**, *112*, 9433-9444.
25. Shirom, M.; Stein, G. *J. Chem. Phys.* **1971**, *55*, 3372-3378.
26. Shirom, M.; Stein, G. *Israel J. Chem.* **1969**, *7*, 405-412.
27. Capone, S.; de Robertis, A.; Sammartano, S.; Rigano, C. *Thermochimica Acta* **1986**, *102*, 1-14.
28. Lenchenkov, V. A.; Vilchiz, V. H.; Bradforth, S. E. *unpublished*.
29. Pommeret, S.; Naskrecki, R.; Meulen, P. v. d.; Menard, M.; Vigneron, G.; Gustavsson, T. *Chem. Phys. Lett.* **1998**, *288*, 883-840.
30. Kee, T. W.; Son, D. H.; Kambhampati, P.; Barbara, P. F. *J. Phys. Chem. A* **2001**, *105*, 8434-8439.

Chapter 9

Near-Threshold Photoionization Dynamics of Indole in Water

Jorge Peon, J. David Hoerner, and Bern Kohler[*]

Department of Chemistry, Ohio State University, 100 West 18th Avenue, Columbus, OH 43210

Abstract

The photoionization of indole in water was studied at 262 nm by the femtosecond transient absorption technique. The initial excited state generated by one-photon absorption is about 0.4 eV above the photoionization threshold of indole in water. Under these near-threshold conditions ionization most likely occurs by an electron-transfer mechanism in which an electron is directly transferred to a trapping site in the solvent without the intermediate generation of a delocalized electron in the conduction band of water. Solvated electrons were formed within our time resolution ($\approx$ 200 fs). No recombination between the solvated electron and the indole radical cation could be observed up to 600 ps after the pump pulse, despite the very high electron affinity of the latter species. A similar result was obtained for aqueous tryptophan. The absence of diffusion-limited charge recombination for this highly exergonic reaction is consistent with Marcus-inverted behavior.

Introduction

Photoionization is a fundamental photochemical decay channel of an electronically excited molecule. When a neutral molecule is photoionized, an electron and a positive ion are produced, and these charged products interact strongly with the molecules in the liquid. Consequently, less energy is needed to photoionize a molecule in a polar solvent compared to vacuum conditions. The liquid environment also profoundly alters the dynamics of photoionization, just as it does for other photochemical processes of excited molecules. These effects are still poorly understood. We have been studying the near-threshold photoionization of aromatic solutes in polar solvents in order to learn more about the microscopic mechanisms underlying charge separation and recombination in the liquid phase.

For an isolated, uncharged molecule, the first ionization potential is the minimum energy required to produce a free electron with zero kinetic energy, which is infinitely far away from its parent ion. This definition is problematic for a molecule in a liquid because of the difficulty of detecting electrons with high efficiency and in an energy-resolved manner in a condensed phase. Electrons can take on a wide range of energies in liquids. In a polar solvent the solvated electron, e_s^-, is the most stable form of the excess electron. The solvated electron in water, e_{aq}^-, lies at least 1.7 eV lower in energy than a delocalized ("quasi-free") electron in the lowest energy state in the conduction band. This fact suggests that it might be possible to photoionize a molecule in a liquid by direct formation of e_s^- without ever forming a conduction-band electron, e_{cb}^-. Such a mechanism requires less energy, and would therefore be particularly important near threshold. Sander, Luther, Troe discussed this issue in an elegant paper that divided liquid-phase photoionization mechanisms into two classes, depending on whether ionization and electron localization are distinct mechanistic steps (1). When these steps are distinct, e_{cb}^- is an intermediate in photoionization, and the mechanism, which we refer to as conduction-band photoionization (CB-PI), can be represented as follows,

$$M^* \quad \overset{\text{ionization}}{\longrightarrow} \quad M^{\bullet+} + e_{cb}^- \quad \overset{\text{localization}}{\longrightarrow} \quad M^{\bullet+} + e_t^- \quad \overset{\text{solvation}}{\longrightarrow} \quad M^{\bullet+} + e_s^- \quad \text{(CB-PI)}$$

In this mechanism, an electron is "ejected" from an electronically excited state, M^*, of molecule M into the conduction band of the liquid. Ionization is thus the initial act of charge separation, which creates the molecular ion, $M^{\bullet+}$. Since charge separation depends primarily on M^*, ionization occurs in essentially the same manner as for an isolated molecule in vacuum. The conduction band electron, e_{cb}^-, may travel a considerable distance before becoming localized at a suitable trapping site in the liquid. The trapped electron, e_t^-, then undergoes further solvation to yield a fully solvated electron, e_s^-. Solvation of $M^{\bullet+}$ also occurs during this time period, but has not been explicitly included in the above scheme.

When ionization and localization are part of the same elementary step, ionization may occur without formation of e_{cb}^-. Instead, an electron is transferred directly to a trapping site. We refer to this type of mechanism as electron-transfer photoionization (ET-PI),

$$M^* \xrightarrow{\begin{array}{c} \text{ionization,} \\ \text{localization} \end{array}} M^{\bullet+} + e_t^- \xrightarrow{\text{solvation}} M^{\bullet+} + e_s^- \qquad \text{(ET-PI)}$$

In this case, ionization occurs due to electronic coupling between M* and the localization site. Because charge separation in ET-PI occurs by transferring an electron from the excited state of the ionizing molecule to an acceptor site, the rate of ionization is likely to depend sensitively on factors that control conventional electron transfer (ET) reactions such as free energies, donor-acceptor geometries, and the rate of solvent fluctuations. Here, the acceptor site could be a single solvent molecule, but is more likely a suitable cluster, which provides a high electron affinity site for an incipient cavity electron. This mechanism, which is impossible for an isolated molecule, can occur for excitation energies below those required for the production of e_{cb}^- and it is therefore also called sub-conduction-band photoionization. Crowell and Bartels have termed this "indirect photoionization" (2). ET-PI has been discussed extensively to explain the photoionization of neat water at low excitation energies (3-5). It has also been discussed in connection with solute photoionization (6).

M* is the immediate precursor to ionization. It is most likely a continuum state in the case of CB-PI, although an autoionizing bound state is also conceivable (1). For ET-PI, M* is always an electronic bound state, although it may not be the same excited state prepared by the optical transition. Recent calculations suggest that the ionizing state of indole is optically dark, but can be accessed by vibronic coupling to optically allowed states (7,8). The rate of ET-PI will then depend on both excited state dynamics of the ionizing molecule and on fluctuations in the environment that promote electron transfer. A finite rate of photoionization has now been observed in ultrafast photodetachment experiments on negative ions in solution (9,10). While it has been proposed that tetramethyl-p-phenylenediamine can be photoionized from its lowest excited singlet (S_1) state (11), photoionization appears to only occur from higher-lying electronic states for most molecules. Higher singlet states generally relax in < 100 fs by internal conversion to S_1. We therefore believe that the very rapid decay of upper singlet states will limit ET-PI to timescales of a few tens to hundreds of femtoseconds for most molecules.

Photoionization in liquids gives rise to a second phenomenon that has no counterpart in isolated molecules: charge recombination. This process, the analog of back electron transfer, refers to the reaction of e_s^- with $M^{\bullet+}$ to reform the parent molecule, $M^{\bullet+} + e_s^- \rightarrow M$. In CB-PI, e_{cb}^- is quite mobile, but still traps within several multiples of the Onsager distance from the initial ionization site. The total distance traveled is thought to depend on the excess energy. In ET-PI, the short timescale for charge separation rules out long-range electron transfer, suggesting that the $M^{\bullet+}$, e_s^- pairs will always be separated by small distances. For both mechanisms, photoionization generates a pair of geminate ions that are separated by some distance. Aided by their mutual Coulombic attraction, they may undergo recombination or diffuse away from each other. The geminate recombination dynamics provide information about the initial ion pair separation. Thus, energy-independent geminate recombination dynamics for water at final state energies below 9 eV provided evidence for a sub-conduction-band photoionization mechanism (2).

Our interest in these issues has led us to study the photoionization of neutral aromatic molecules in polar solvents using femtosecond pump-probe methods. Charge ejection following absorption of a UV or near UV photon is a common deactivation channel for many aromatic compounds in solution (12). This allows these solutes to be monophotonically ionized under carefully controlled excitation conditions, circumventing some of the difficulties in multiphoton ionization experiments on molecules such as water. We are concentrating on the near-threshold regime where ET-PI is expected to dominate. In addition to looking for 'delayed photoionization' due to this mechanism, an additional motivation was to quantify the actual photoionization yield at short times before geminate recombination has reduced the yield of ions. The geminate recombination dynamics provide insight into reactions of a simple radical ion pair consisting of an organic radical cation and the solvated electron. Unlike neat water, geminate recombination involves reaction between two partners instead of three. The rates of reaction between solvated electrons and a large number of substrates vary over a wide range, but are often masked by diffusion. Tight ion pairs created by ET-PI make it possible to investigate true reaction rates in the absence of transport.

We chose indole for these first studies of ultrafast solute photoionization for several reasons. Indole photoprocesses have been extensively studied for decades due to the importance of this molecule as the chromophore of the aromatic amino acid, tryptophan. The exquisite sensitivity of the indole chromophore to its local environment has been used extensively to study proteins (13). This important application has been the stimulus for hundreds of studies of indole photoprocesses. This past work has identified photoionization as an important deactivation channel for electronically excited indole, but the mechanism is still unclear. Indole is also one of the only polyatomic molecules for which condensed-phase photoionization thresholds have been measured in a range of solvents. The data from Bernas and co-workers (14) is summarized in Table 1.

Table I. Photoionization thresholds for indole in various solvents.[a]

Solvent	Threshold Energy / eV
1-butanol	5.40[b]
1-propanol	5.15[b]
tetramethylsilane	4.95
ethanol	4.85
methanol	4.60
H_2O	4.35

[a]Data from ref 14.

[b]Extrapolated from values for shorter alcohols.

Results and Discussion

In our experiments (*15*), the third harmonic pulse ($\tau \approx 180$ fs) from a femtosecond titanium:sapphire laser was used to excite a 6 mM aqueous solution of indole at $\lambda = 262$ nm (see Figure 1). A continuum probe pulse was then used to measure transient absorption at wavelengths throughout the visible and near-IR. Figure 2 compares the transient signals for aqueous indole (top panel) with ones obtained from the two-photon ionization of neat water using the same pump and probe wavelengths (lower panel). The signals for aqueous indole were recorded at low pump intensities. They result solely from monophotonic

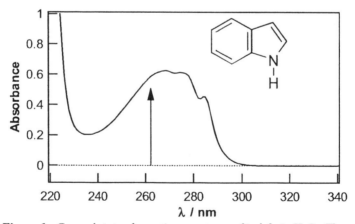

Figure 1. *Ground state absorption spectrum of indole in H_2O. The arrow shows the pump wavelength used for the femtosecond transient absorption experiments.*

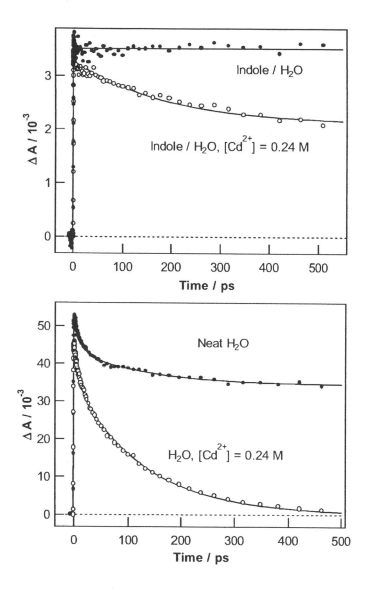

Figure 2. *Transient absorption at a probe wavelength of 700 nm for a 6 mM solution of indole in H₂O (top), compared with signals from two-photon ionization of neat water (bottom). The pump wavelength in both cases is 262 nm.*

excitation of indole. To monitor electron formation and decay, probing for all scans in Figure 2 was done at 700 nm, near the absorption maximum of e_{aq}^-. The water cation decays in less than 100 fs by proton transfer to a neighboring water molecule to give $OH^•$ and H_3O^+. These species absorb only in the UV and therefore do not contribute to the signal at 700 nm. The distinctive, nonexponential decay seen in neat water (figure 2, lower plot) is a hallmark of geminate recombination of e_{aq}^- with $OH^•$ and H_3O^+ (2). Adding cadmium perchlorate provides definitive proof of ionization. This salt does not change the absorbance of the neat water sample at the pump wavelength, but has a profound effect on the dynamics as shown in the lower plot of Figure 2. Cd^{2+} is a potent scavenger that reacts with e_{aq}^- near the diffusion-limited rate. The decay of e_{aq}^- in water with 0.24 M Cd^{2+} can be fit to a single exponential with a time constant of 126 ps, corresponding to a bimolecular reaction rate of 3.3×10^{10} M^{-1} s^{-1}. This compares reasonably well with the value of $\approx 4 \times 10^{10}$ M^{-1} s^{-1} from scavenging experiments that used cadmium perchlorate concentrations between 0.1 and 0.5 M (16).

Most striking about the indole / water signal (Figure 2, upper plot) is the absence of any appreciable decay up to 600 ps after the pump pulse. For the small electron-cation separations expected near the photoionization threshold, rapid geminate recombination between $Ind^{•+}$ and e_{aq}^- should produce a noticeable loss of e_{aq}^- within our time window. Its absence suggests that recombination does not occur at every encounter, as discussed in more detail below. Unlike the neat water case, photoionization of indole produces multiple species that contribute to the induced absorbance at the probe wavelength. One indole radical cation ($Ind^{•+}$) will be formed for every photoejected electron. Since the quantum yield for photoionization of indole is small at our pump wavelength, a large number of indole molecules will be present in the S_1 (= ^1L_a) state. The decay time of the S_1 population is equal to the fluorescence lifetime, which for indole in water is nearly 5 ns (17). Excited state absorption ($S_1 \rightarrow S_N$ absorption) by this population will create a very slowly decaying signal offset on the subnanosecond timescale of interest here.

Several experiments were undertaken to analyze the relative signal contributions from $Ind^{•+}$ and indole $S_1 \rightarrow S_N$ absorption as a function of the probe wavelength (15). The $Ind^{•+}$ absorption spectrum was measured after generating this radical cation by electron transfer quenching of the indole S_1 state in the presence of CCl_4. The results, which are in excellent agreement with other reports (18,19), indicate little absorption by $Ind^{•+}$ for $\lambda \geq 700$ nm. The indole $S_1 \rightarrow S_N$ absorption spectrum was characterized in 1-propanol in the absence of CCl_4. In this solvent, the photoionization threshold (see Table 1) lies above our photon energy, and only $S_1 \rightarrow S_N$ absorption is observed. As Figure 3 shows, the signal at 700 nm has decayed by less than ten percent 0.5 ns after the pump pulse, consistent with

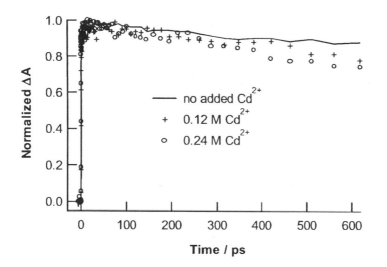

Figure 3. *Transient absorption at 700 nm for indole in 1-propanol as a function of cadmium perchlorate concentration.*

the $\approx$ 4 ns fluorescence lifetime of indole in simple alcohols (*20*). Thus, on the timescale of these experiments, the signature of excited state absorption is a slowly decaying offset.

The addition of Cd^{2+} to indole in 1-propanol (crosses and open circles in Figure 3) produces only modest increases in the signal decay rate, consistent with the absence of photoionization in this solvent. Although the rate of reaction between the excess electron and Cd^{2+} has not been reported in 1-propanol, we assume that it does not differ significantly from its value in water. This is in keeping with the small solvent effects which have been observed for fast scavenging reactions of excess electrons (*21*). The small decay that is observed at 700 nm is consistent with a minor quenching of the S_1 state. Mialocq et al. reported that the fluorescence intensity of indole is quenched by $\approx$ 20% in the presence of 0.24 M Cd^{2+} (*22*). The small changes observed in Figure 4 are consistent with a decrease in the S_1 lifetime of this magnitude. This effect may originate in enhanced intersystem crossing induced by cadmium (*22*).

The pronounced decay observed when cadmium perchlorate is added to aqueous indole (open circles in top panel of Figure 2) indicates that solvated electrons are formed in significant numbers. The signal at 700 nm is fit by a single exponential with a decay time of 184 ps plus an offset. The amplitude of the offset is 63% of the maximum signal amplitude. The nonscavengable signal modeled by the offset is due to $S_1 \rightarrow S_N$ absorption by non-ionized indole molecules. Thus, even though the quantum yield for photoionization is much less than unity, absorption by e_{aq}^- accounts for nearly 40% of the transient signal since its absorption coefficient at 700 nm is nearly five times larger than that of the indole S_1 state. We previously estimated the molar absorption coefficient of the indole S_1 state to be approximately 4 000 M^{-1} cm^{-1} (15). Using this value and the 37% contribution of e_{aq}^- absorption to the 700 nm signal, the photoionization quantum yield, ϕ_{e^-}, is estimated to be 0.11. This value agrees well with $\phi_{e^-} =$ 0.15 by Amouyal et al (23), providing further evidence that the scavengeable signal at 700 nm from indole / water is due to e_{aq}^-. From the decay time of 184 ps, the rate of reaction between e_{aq}^- and Cd^{2+} is nearly 50% slower than when the same concentration of cadmium ions was added to neat water. More work is needed to determine the reason for this decrease. One possibility is that e_{aq}^- generated by indole photoionization is less mobile due to the formation of a long-lived ion pair with the indole radical cation. Alternatively, the close proximity of $Ind^{\bullet+}$ and its associated solvent shell could sterically hinder access to e_{aq}^- by a scavenger such as Cd^{2+}.

For indole in water in the absence of Cd^{2+}, the transient absorption signal at 700 nm shows minimal decay for at least 600 ps after the pump pulse. Flat transient absorption signals were also observed for aqueous solutions of tryptophan as shown in Figure 4 (24). In order to better understand these results, the geminate recombination dynamics were modeled numerically. In general, the motions of the two partners of a geminate ion pair are controlled partially by diffusion, and partially by their mutual Coulombic attraction. The two partners

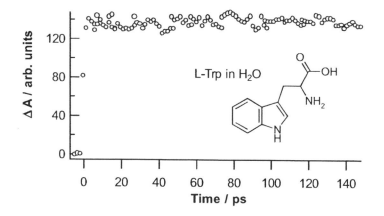

Figure 4. *Transient absorption at 570 nm for L-trytophan in water following excitation with a 180 fs, 262 nm pump pulse.*

of some geminate ion pairs will approach each other closely through diffusion, while others will diffuse apart into the bulk of the liquid, becoming separated by such a great distance that the probability of recombination becomes vanishingly small. A standard approach is to assume that reaction occurs instantaneously whenever the two partners ($Ind^{\bullet+}$ and e_{aq}^{-}) approach one another within a critical reaction distance, R. R should be close to the sum of the radii of approximately spherical reactants. An exact, analytical solution is not available for the time-dependent recombination of ions, so we used an approximate expression appropriate for high dielectric constant liquids such as water (25). We used a value of R = 5 Å, and a value of 4.8×10^{-9} m^2 s^{-1} was used for the diffusion coefficient of e_{aq}^{-}. The diffusion coefficient of $Ind^{\bullet+}$ has not been reported, so we assumed a value of 0.9×10^{-9} m^2 s^{-1} based on the diffusion behavior of charged, organic molecules of similar size. The simulation results are relatively insensitive to the precise values. The final parameter for the simulations was the initial separation, r_0, between e_{aq}^{-} and $Ind^{\bullet+}$.

Simulations for three values of r_0 are shown by the solid curves in Figure 5. Significant recombination is predicted for $r_0 = 10$ Å. This value was chosen because it is believed to be the average separation between e_{aq}^{-} and the initial ionization site when water is photoionized at final state energies below about 9 eV, corresponding to ET-PI. The simulation for $r_0 = 6$ Å shows what is expected for diffusion-limited recombination of tighter ion pairs. In this case, the ultimate escape yield is lower, and recombination should cause an obvious, large amplitude change to the transient absorption signal, particularly at delay times below 50 ps. Only for values of $r_0 \geq 35$ Å are curves obtained that match the flat kinetics observed by us. Since our results show that solvated electrons are produced on a subpicosecond timescale, making long-range ET impossible, such large distances are incompatible with ET-PI. In the CB-PI mechanism, the quasi-free electrons are more mobile and can travel further from the site of ionization. However, the distance traveled should depend sensitively on the excess energy of the ejected electron. If the 4.35 eV threshold reported by Bernas defines the conduction band edge, then the electrons in our experiments would have only $\approx$ 0.4 eV of excess energy. This value is too small to produce $r_0 \geq 35$ Å. In water, excess electron energies of about 3 eV are required to produce such large radius geminate ion pairs (2).

The most reasonable explanation for the absence of recombination at times less than 500 ps is that reaction between e_{aq}^{-} and $Ind^{\bullet+}$ is not diffusion-limited. Perhaps due to steric reasons, or due to the magnitude of electronic coupling between them, $Ind^{\bullet+}$ and e_{aq}^{-} do not react with unit probability on every encounter. In this case, the time-dependent survival probability can be calculated using a model due to Green (26). In this model, a Collins-Kimball reaction velocity is used to account for the fact that reaction does not occur as soon as $Ind^{\bullet+}$ and e_{aq}^{-} are within a distance R of one another. The results are shown by

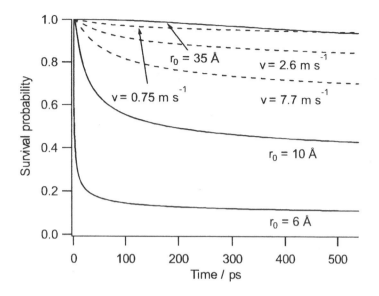

Figure 5. *Simulations of the survival probability of geminate ion pairs created by photoionization of indole in aqueous solution. The solid curves show the approximate solution for diffusion-limited recombination of oppositely charged ions. The dashed curves include a reaction velocity to model non-diffusion-limited recombination.*

the dashed curves in Figure 5, using the same values for the diffusion constants and R as before, and a value of $r_0 = 10$ Å. We believe that this value is an upper limit for the size of the ion pairs produced by near-threshold photoionization. Moving from top to bottom, the indicated reaction velocities correspond to reaction probabilities per encounter of approximately 3%, 10%, and 25%. Clearly, to model slow recombination, it is necessary to assume that reaction between e_{aq}^- and $Ind^{•+}$ is relatively inefficient. The estimate of 3% is likely to be an upper limit based on our choice of r_0 and the fact that we have not measured the actual rate of recombination. Since Mialocq and co-workers found no evidence of recombination between 50 ps and 2 ns in their experiments on aqueous indole with 25 ps pulses (*22,23*), the actual probability of reaction is likely to be even lower.

Very few reaction rates between e_{aq}^- and open-shell species are known. An important exception is the rate of scavenging of e_{aq}^- by the hydroxyl radical. In the absence of more extensive data, it is difficult to predict the rate of reaction with $Ind^{•+}$. In fact, predicting the rates of reaction for e_{aq}^- with closed-shell scavengers is still very difficult. In this case, the observed rate constants vary over 10 orders of magnitude (27). Nonetheless, faster rates are generally observed for substrates with greater gas-phase electron affinities (27). The electron affinity of Ind$^{\bullet+}$ in the gas phase is equal to the ionization potential of neutral indole (7.76 eV (28)). This value greatly exceeds the electron affinity of indole, which was recently estimated to be –1.03 V from cluster studies (29). On this basis, we expected scavenging of e_{aq}^- by Ind$^{\bullet+}$ to occur at a much higher rate than scavenging by neutral indole ($k \approx 2 \times 10^8$ M^{-1} s^{-1} (30)).

Geminate recombination between the products of photoionization is a back electron transfer reaction and should be amenable to treatment using Marcus theory (31). This has been done only rarely for reactions involving e_s^-. In one such study, evidence was presented for Marcus-inverted behavior at high exergonicities in THF (32). It is plausible that the Ind$^{\bullet+}$ and e_{aq}^- reaction is also in the Marcus-inverted region and therefore reacts slower than expected. Using the oxidation potential of indole (1.24 V (18)), and the reduction potential of the solvated electron (–2.9 V (33)), the driving force for back ET is $\approx$ –4.1 V (vs. NHE). This is considerably more negative than in most back ET reactions. We note that slow geminate recombination due to a large, negative free energy change was observed in simulations of aqueous halide ions (34,35). Further experiments are needed to explore how Marcus theory and its contemporary variations can be applied to radical ion-solvated electron recombination in liquids.

Although our observation of slow geminate recombination at a single excitation energy cannot be used to decide between ET-PI or CB-PI as the mechanism for solvated electron formation, we favor the former mechanism for aqueous indole. The same view was taken by Kevan and Steen who proposed in 1975 that indole photoionization directly generates solvated electrons without an intermediate conduction band electron (36). A recent reassessment (37) of the conduction band edge of water makes it difficult to reconcile the threshold value of 4.35 eV with promotion of an electron into the conduction band.

Recent ab initio calculations by Sobolewski and Domcke also provide support for ET-PI in indole (7,8). These authors have identified a $^1\pi\sigma^*$ state which lies slightly higher in energy than the ^1L_a and ^1L_b states. Since transitions to the $^1\pi\sigma^*$ state are optically forbidden, it can only be accessed via internal conversion from the ^1L_a and ^1L_b states through vibronic coupling. This highly polar excited state (permanent dipole moment $\approx$ 10 D) has a diffuse electron density distribution centered on the hydrogen atom attached to nitrogen, giving it considerable Rydberg character. The $^1\pi\sigma^*$ state predissociates the ^1L_a and ^1L_b states along the N-H stretch coordinate. For isolated indole a conical intersection with S_0 is found at large N-H bond distance, which provides a mechanism for nonradiative return to the electronic ground state (7). For indole clustered with 1-3 water molecules, the diffuse electron cloud of the $^1\pi\sigma^*$ state is captured by the water molecules, giving the precursor of the solvated electron.

In support of their interpretation, they note that the adiabatic excitation energy of the $^1\pi\sigma^*$ state appears to converge to an energy close to the observed photoionization threshold for indole in water (4.35 eV, Table I). This ab initio study shows in detail how an upper electronic state can transfer an electron directly to the solvent.

Conclusions

We have investigated the ultrafast dynamics of photoexcited indole in aqueous solution using femtosecond pump-probe spectroscopy. This work provides new insight into the photophysics and photochemistry of indole. Our measurements show that photoionization is a significant decay channel at 262 nm. Electron ejection occurs within our instrumental time resolution of $\approx$ 200 fs, consistent with electron ejection from an excited state above S_1. Excited singlet state absorption by indole and the absorption spectrum of the indole radical cation were characterized to determine the extent to which they mask the dynamics of e_{aq}^-. There is minimal recombination between 0 and 600 ps after photoionization. In view of the small radius geminate ion pairs that are generated by near-threshold ionization, we conclude that the bimolecular rate of reaction between e_{aq}^- and the indole radical cation is substantially slower than the diffusion limit. It is not currently known if the electron and indole radical cation exist in a tight ion pair for any appreciable time, or if they quickly diffuse away from one another. Finally, highly exergonic charge recombination between e_{aq}^- and a molecular radical cation may be another example of Marcus-inverted behavior in liquids.

References

1. Sander, M. U.; Luther, K.; Troe, J. *Ber. Bunsenges. Phys. Chem.* **1993**, *97*, 953.
2. Crowell, R. A.; Bartels, D. M. *J. Phys. Chem.* **1996**, *100*, 17940.
3. Goulet, T.; Bernas, A.; Ferradini, C.; Jay-Gerin, J.-P. *Chem. Phys. Lett.* **1990**, *170*, 492.
4. Bernas, A.; Grand, D. *J. Phys. Chem.* **1994**, *98*, 3440.
5. Keszei, E.; Jay-Gerin, J.-P. *Can. J. Chem.* **1992**, *70*, 21.
6. Steen, H. B.; Bowman, M. K.; Kevan, L. *J. Phys. Chem.* **1976**, *80*, 482.
7. Sobolewski, A. L.; Domcke, W. *Chem. Phys. Lett.* **1999**, *315*, 293.
8. Sobolewski, A. L.; Domcke, W. *Chem. Phys. Lett.* **2000**, *329*, 130.
9. See Kloepfer, J. A.; Vilchiz, V. H.; Lenchenkov, V. A.; Bradforth, S. E. elsewhere in this volume.

10. Barthel, E. R.; Martini, I. B.; Schwartz, B. J. *J. Chem. Phys.* **2000**, *112*, 9433.
11. Hirata, Y.; Mataga, N. *J. Phys. Chem.* **1983**, *87*, 3190.
12. Joschek, H.-I.; Grossweiner, L. I. *J. Am. Chem. Soc.* **1966**, *88*, 3261.
13. Callis, P. R. In *Methods in Enzymology*; Academic Press: New York, 1997; Vol. 278, pp 113.
14. Bernas, A.; Grand, D.; Amouyal, E. *J. Phys. Chem.* **1980**, *84*, 1259.
15. Peon, J.; Hess, G. C.; Pecourt, J.-M. L.; Yuzawa, T.; Kohler, B. *J. Phys. Chem. A* **1999**, *103*, 2460.
16. Aldrich, J. E.; Bronskill, M. J.; Wolff, R. K.; Hunt, J. W. *J. Chem. Phys.* **1971**, *55*, 530.
17. Ricci, R. W. *Photochem. Photobiol.* **1970**, *12*, 67.
18. Shen, X.; Lind, J.; Merenyi, G. *J. Phys. Chem.* **1987**, *91*, 4403.
19. Jovanovic, S. V.; Steenken, S. *J. Phys. Chem.* **1992**, *96*, 6674.
20. Steiner, R. F.; KIrby, E. P. *J. Phys. Chem.* **1969**, *73*, 4130.
21. Bolton, G. L.; Freeman, G. R. *J. Am. Chem. Soc.* **1976**, *98*, 6825.
22. Mialocq, J. C.; Amouyal, E.; Bernas, A.; Grand, D. *J. Phys. Chem.* **1982**, *86*, 3173.
23. Amouyal, E.; Bernas, A.; Grand, D.; Mialocq, J. C. *Faraday Discuss. Chem. Soc.* **1982**, 147.
24. Peon, J.; Kohler, B. *Unpublished data.*
25. Clifford, P.; Green, N. J. B.; Pilling, M. J. *J. Phys. Chem.* **1984**, *88*, 4171.
26. Green, N. J. B. *Chem. Phys. Lett.* **1984**, *107*, 485.
27. Hart, E. J.; Anbar, M. *The hydrated electron*; Wiley-Interscience: New York, 1970.
28. Hager, J. W.; Wallace, S. C. *Anal. Chem.* **1988**, *60*, 5.
29. Carles, S.; Desfrancois, C.; Schermann, J. P.; Smith, D. M. A.; Adamowicz, L. *J. Chem. Phys.* **2000**, *112*, 3726.
30. Armstrong, R. C.; Swallow, A. J. *Radiat. Res.* **1969**, *40*, 563.
31. Marcus, R. A. *J. Chem. Phys.* **1965**, *43*, 3477.
32. Kadhum; Salmon *J. Chem. Soc. Faraday Trans. I* **1986**, *82*, 2521.
33. Schwarz, H. A. *J. Chem. Educ.* **1981**, *58*, 101.
34. Sheu, W.-S.; Rossky, P. J. *J. Phys. Chem.* **1996**, *100*, 1295.
35. Staib, A.; Borgis, D. *J. Chem. Phys.* **1996**, *104*, 9027.
36. Kevan, L.; Steen, H. B. *Chem. Phys. Lett.* **1975**, *34*, 184.
37. Coe, J. V.; Earhart, A. D.; Cohen, M. H.; Hoffman, G. J.; Sarkas, H. W.; Bowen, K. H. *J. Chem. Phys.* **1997**, *107*, 6023.

Chapter 10

Intermolecular Hydrogen Bonding in Chlorine Dioxide Photochemistry: A Time-Resolved Resonance Raman Study

Matthew P. Philpott[1], Sophia C. Hayes[1], Carsten L. Thomsen[2], and Philip J. Reid[1,*]

[1]Department of Chemistry, University of Washington, Box 351700, Seattle, WA 98195–1700
Current address: Department of Chemistry, University of Åarhus, Langelandsgade 140, DK–8000 Åarhus C, Denmark

The geminate recombination and vibrational relaxation dynamics of chlorine dioxide (OClO) dissolved in ethanol and 2,2,2-trifluoroethanol (TFE) are investigated using time-resolved resonance Raman spectroscopy. Stokes spectra are measured as a function of time following photoexcitation using 398-nm degenerate pump and probe wavelengths. In ethanol, subpicosecond reformation of ground-state OClO through recombination of the primary photofragments occurs with a quantum yield of 0.5 ± 0.1. Following recombination, intermolecular vibrational relaxation occurs with a time constant of 31 ± 10 ps. For OClO dissolved in TFE, recombination occurs with a time constant of 1.8 ± 0.8 ps and quantum yield of only 0.3 ± 0.1. In addition, intermolecular vibrational relaxation occurs with a time constant of 79 ± 27 ps. The decreased geminate-recombination quantum yield and vibrational-relaxation rate for OClO dissolved in TFE is interpreted in terms of increased self-association of this solvent relative to ethanol.

Introduction

Determining how solvents influence geminate-recombination and vibrational-relaxation dynamics is central to understanding chemical reactivity in condensed media.[1-3] The majority of work in this area has focused on the condensed-phase dynamics of diatomics. Photodisoociation of I_2 in the condensed phase is followed by recombination of the nascent photofragments resulting in the formation of vibrationally-excited I_2 which undergoes subsequent relaxation through intermolecular vibrational relaxation.[4-15] The efficiency of geminate recombination is dependent on the initial collisional dynamics between the photofragments and surrounding solvent molecules (i.e., the solvent cage). The subpicosecond geminate recombination rate observed for I_2 is similar to the behavior observed in other systems suggesting that the rate of recombination is relatively insensitive to the details of the solvent.[5,16] Ground-state vibrational relaxation of I_2 proceeds on the tens-to-hundreds of picoseconds timescale depending on solvent.[4,5] The relaxation rate has been modeled using isolated-binary-collision theory consistent with short-range forces dominating the solvent-solute interaction dynamics.[5]

A more complex picture has emerged for charged and/or polar diatomics in solution where electrostatic as well as collisional solvent-solute coupling is operative. This complexity is well illustrated by recent studies of I_2^-.[17-21] Here, the geminate-recombination quantum yield in water was found to be two and four times greater than in ethanol or acetonitrile, respectively.[18] The specific solvent-solute interactions responsible for this variation are not well understood at present. The vibrational-relaxation time constant for I_2^- increases from 2.8 ps in water to 4.5 ps in ethanol or acetonitrile, with this relaxation remarkably accelerated relative to I_2.[17,18,21-24] Through comparison with MD simulations, this acceleration has been assigned to electrostatic solvent-solute interactions that are enhanced by "charge-flow" accompanying vibrational motion.[19,20] Experimental and theoretical studies of other charged an/or polar systems have also found vibrational relaxation times that are shorter than expected for purely collisional dynamics, further supporting the importance of electrostatic interactions.[25-29]

Recently, there has been a great deal of interest in extending the description of geminate recombination and vibrational relaxation developed from studies of diatomic systems to triatomics. Barbara and coworkers have observed slow geminate-recombination (3.5 ps) and extremely fast vibrational-relaxation ($<<3.5$ ps) for aqueous O_3^-, and have attributed these effects to strong electrostatic solvent-solute interaction.[30] Studies by Hochstrasser and coworkers have found that the vibrational relaxation time constant for N_3^- dissolved in water, deuterated water, methanol, or ethanol is <3 ps.[31] In contrast, the vibrational relaxation time-constant in polar, aprotic hexamethyl-phosphamide is ~15 ps consistent with intermolecular hydrogen bonding accelerating the rate of vibrational relaxation.[31] Finally, recent studies by Keiding and coworkers on CS_2 have demonstrated that the vibrational relaxation of CS$_2$ in water occurs with two vibrational relaxation times of ~8 and ~33 ps, substantially slower than the times observed for charged triatomics in solution.[32]

Chlorine dioxide (OClO) represents an excellent opportunity to extend to our understanding of triatomic reactivity in condensed environments. OClO is a 19-electron neutral radical that is isoelectronic with O$_3^-$. The $^2B_1$ (X)-$^2A_2$ (A) electronic transition of OClO is centered at ~360 nm.[33,34] The $^2A_2$ surface is predissociative, with internal conversion to the $^2A_1$ and subsequent decay of this surface to the $^2B_2$ state resulting in dissociation to produce ClO and O, or Cl and O$_2$.[34] Recent femtosecond pump-probe and time-resolved resonance Raman (TRRR) studies have demonstrated that sub-picosecond geminate recombination results in the reformation of vibrationally-excited ground-state OClO.[35-44] To date, the geminate recombination and vibrational relaxation dynamics of OClO have only been studies in two solvents, water and acetonitrile. Both the geminate-recombination quantum yield and intermolecular-vibrational-relaxation rate are substantially reduced in acetonitrile relative to water.[40,42] This observation has lead to the suggestion that intermolecular hydrogen-bonding makes a substantial contribution to solvent-solute coupling.

In this manuscript, we present a series of studies designed to explore the role of intermolecular hydrogen bonding in defining the geminate-recombination and vibrational-relaxation dynamics of OClO. Subpicosecond TRRR and studies are performed on OClO dissolved in water, ethanol, and 2,2,2-trifluoroethanol (TFE). In theory, the solvent series investigated here when combined with our previous results on aqueous OClO provides the opportunity to investigate the role of intermolecular hydrogen bonding in the reaction dynamics of OClO. In ethanol, subpicosecond geminate recombination of the primary photoproducts is observed, followed by vibrational relaxation of OClO with a time constant of ~30 ps. In contrast, the time-constant for geminate recombination is ~1.8 ps in TFE, significantly delayed relative to any other solvent studied to date. In addition, the geminate-recombination quantum yield is 0.5 in ethanol, but decreases to 0.3 in TFE. Both quantum yields are substantially reduced relative to the 0.9 quantum yield observed in water. The vibrational-relaxation time constant for OClO dissolved in TFE is ~80 ps, significantly longer compared to water or ethanol. The reduced geminate-recombination quantum yield, delayed recombination, and slower vibrational relaxation in TFE is assigned to greater self-association of the solvent as well as stabilization of the photofragments through solvent complexation.

Materials and Methods

The laser spectrometer has been described in detail elsewhere.[40] Briefly, a home-built Ti:Sapphire oscillator pumped by an argon-ion laser (operating all-lines) produced 30-fs pulses (full-width at half-maximum) centered at 795 nm at a repetition rate of 91 MHz. The pulses were amplified by a Ti:sapphire regenerative amplifier (Clark-MXR CPA-1000-PS). The amplified output consisted of 700-µJ pulses, tunable from 770 to 810 nm, at a repetition rate of 1 kHz. The 398-nm pump and probe beams were generated by frequency doubling the 795-nm amplifier output in a 1-mm thick β-BBO crystal (Type I). The polarization of the pump was rotated to 54.7° relative to the probe using a zero-order half-wave plate to minimize the contribution of rotational dynamics to the data. Pulse energies of 4 and 0.9 µJ were employed for the pump and probe, respectively. A 135° backscattering geometry was employed, with the beams focused onto a 2-mm path length fused-silica flow cell containing ~30 mM OClO solutions prepared as described elsewhere.[45] Pump and probe overlap was optimized by monitoring the pump-induced absorption using a photodiode located behind the sample. The scattered light was collected with standard, refractive UV-quality optics and delivered to a 0.5-m focal-length spectrograph (Acton 505F) equipped with a 2400-grooves/mm holographically-ruled grating. Spectrometer slit widths were adjusted to provide ~15 cm^{-1} resolution. The scattered light was detected by a liquid nitrogen-cooled, 1340 x 100 pixel, back-thinned, CCD detector (Princeton Instruments). The instrument response as measured by the optical Kerr effect in water was 1.2 ± 0.1 ps. The time-resolved difference spectra presented here were constructed as follows. At each time delay, three Raman spectra were obtained with the "probe only", "pump and probe", and the "pump only" incident on the sample. Six-minute integration times were employed for each spectrum. The pump-only spectrum was subtracted from the pump-and-probe to produce the "probe-with-photolysis" spectrum. The probe-only spectrum then was subtracted from the probe-with-photolysis spectrum to produce the difference spectra reported here. All intensities were determined to scale linearly with pump and probe intensity. The time-dependent evolution in scattered intensity was analyzed by fitting the data to a convolution of the instrument response with a sum of exponentials using the Levenberg-Marquardt algorithm. A minimal number of exponentials were employed to produce an accurate fit, with goodness of fit determined by the reduced χ^2 values and visual inspection of the residuals. Error values for represent one standard deviation from the mean. Infrared absorption spectra were obtained on a FTIR spectrometer (Perkin Elmer 1720) at 1-cm^{-1} resolution. Samples were placed into a NaCl windowed cell having a 0.5-mm path length.

Results

OClO in Ethanol

Figure 1 presents the time-resolved resonance Raman Stokes difference spectra of OClO in ethanol. The 0-ps difference spectrum demonstrates substantial scattering depletion for the OClO symmetric-stretch fundamental and overtone transitions. As the delay between the pump and probe increases, the extent of OClO depletion decreases up to ~80 ps. Comparison of the initial depletion amplitude to the depletion that persists at later delays establishes that the geminate recombination quantum yield in ethanol is 0.5 ± 0.1, substantially reduced relative to water (0.90 ± 0.10).[42]

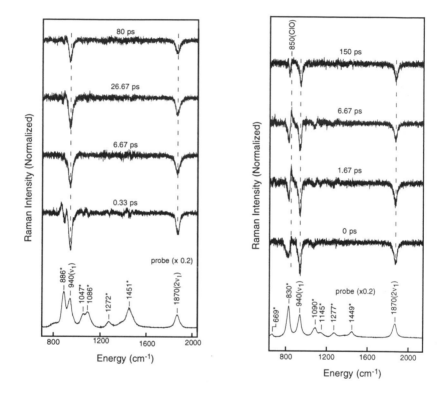

Figure 1. Time-resolved resonance Raman Stokes difference spectra of OClO dissolved in ethanol (left) and 2,2,2-trifluoroethanol (right). Pump and probe wavelengths are both 398 nm. The time delay between the pump and probe for a given spectrum is indicated. The probe-only spectrum is presented at the bottom of each panel. Transitions corresponding to the solvent are marked with an asterisk in the probe-only spectrum.

To avoid interpretive difficulties due to overlap between the OClO symmetric-stretch fundamental transition and the 886 cm^{-1} solvent transition, kinetic analysis was performed using the intensity of the symmetric-stretch overtone transition at 1870 cm^{-1} as presented in Figure 2A. Inspection of the temporal evolution in depletion intensity demonstrates that the depletion recovery is biphasic. Consistent with this observation, the data were best modeled by a sum of three exponentials convolved with the instrument response resulting in recovery time-constants of 0.3 ± 0.1 ps (instrument-response limited), 31 ± 10 ps, and a fixed offset representing the long-time scattering depletion. The instrument response limited reformation of ground-state OClO is similar to behavior observed in water and acetonitrile.[42] In addition, the later time-recovery has been shown to correspond to the vibrational relaxation of OClO.[42] Compared to the 9-ps relaxation time observed in water, the ~30-ps time observed in ethanol demonstrates that the intermolecular vibrational-relaxation rate is substantially reduced in this solvent.

OClO in 2,2,2-Trifluoroethanol (TFE)

Figure 1 also presents time-resolved resonance Raman Stokes difference spectra of OClO dissolved in TFE. At 0-ps delay, negative intensity is observed for transitions corresponding to OClO consistent with ground-state depletion created by photolysis. In addition, negative intensity is observed for the 830-cm^{-1} transition of TFE implying that the optical density of the sample has increased; however, pump-probe studies (data not presented) have shown that the optical density decreases at early times (<4 ps). In addition, negative solvent intensity persists for delays outside the pump-probe temporal overlap inconsistent with pump-pulse induced frequency modulation of the solvent transition. We will argue below that this intensity arises from strong interaction between OClO and the solvent. The depletion in OClO scattering intensity decreases for delays up to 150 ps. Figure 2C displays the integrated intensity of the OClO symmetric-stretch overtone transition as a function of pump-probe delay. The residual depletion observed at longer delays relative to the initial depletion is consistent with a geminate-recombination quantum yield of 0.3 ± 0.1 in TFE. Best fit to the data was accomplished by a sum of three exponentials convolved with the instrument response resulting in recovery time-constants of 1.8 ± 0.8 ps, 79 ± 27 ps, and a fixed offset representing the long-time scattering depletion. The ~80-ps recovery time constant demonstrates that the intermolecular-vibrational-relaxation rate in TFE is significantly reduced relative to water or ethanol. In addition, the early-time recovery time-constant of 1.8 ps is substantially longer than the instrument-response limited recovery observed in all other solvents studied to date suggesting that the geminate recombination is slower in this solvent. However, the limited number of data points obtained in this time region requires critical evaluation of this result. Figure 2B presents a comparison between the best fit to the ethanol data as described above and a fit were the early-time
recovery is fixed to be 1.8 ps. For this later fit, χ^2 is ~8 times larger than that for the best fit. Figure 2D presents a comparison between the best fit to the TFE

142

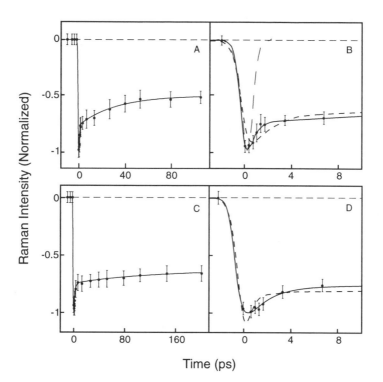

Figure 2. (A) Intensity of the OClO symmetric-stretch overtone transition as a function of time for OClO in ethanol. Best fit to the data by the sum of three exponentials convolved with the instrument response (solid line) resulted in time constants of 0.3 ± 0.1 ps (normalized amplitude of -0.82 ± 0.05), 31 ± 10 ps (-0.06 ± 0.03), and a long-time offset (-0.12 ± 0.04). (B) Expanded view of the early time recovery in scattered intensity presented in panel A. In addition to the best fit described above (solid line) a second fit obtained with the shortest time-constant is held to 1.8 ps is also presented (dashed line). The instrument response is given by the long-dashed line centered at 0-ps delay. (C) Intensity of the OClO symmetric-stretch Stokes overtone transition in 2,2,2-trifluoroethanol as a function of time. Best fit to the data by the sum of three exponentials convolved with the instrument response (solid line) resulted in time constants of 1.8 ± 0.8 ps (normalized amplitude of -0.39 ± 0.08), 31 ± 10 ps (-0.07 ± 0.04), and long-time offset (-0.54 ± 0.08). (D) Expanded view of the early time recovery in Stokes intensity presented in panel C. In addition to the best fit described above (solid line) a second fit obtained with the shortest time-constant is held to 0.3 ps is also presented (dashed line).

data and a fit where the early-time constant is fixed to 0.3 ps. Compared to the best fit, χ^2 increases by an order of magnitude. The comparative analyses presented in Figure 2B and 2D suggest that the rate of geminate recombination is reduced in TFE. Support for this result is also found in Figure 1 where positive intensity is observed at 850 cm^{-1} consistent with ClO formation. Analysis of the ClO scattering intensity is not possible due to overlap with the solvent scattering; however, the presence of positive intensity in this frequency region suggests that ClO is indeed produced following OClO photoexcitation

Discussion

Geminate Recombination

The TRRR data presented here provide limited support for a correlation between geminate recombination efficiency and intermolecular hydrogen bonding. The geminate-recombination quantum yields in ethanol (~0.5) and TFE (~0.3) are substantially reduced compared to water (~0.9) consistent with the modulation of intermolecular hydrogen bonding in these solvents relative to water. Quantitative comparison of intermolecular hydrogen bonding strengths can be performed using the Kalmut-Taft α-scale.[46] The α parameter is one of a collection of solvochromatic parameters that correlates specifically with the hydrogen-bond donating strength of the solvent. According to this scale, water ($\alpha = 1.17$) is a better hydrogen-bond-donating solvent than ethanol ($\alpha = 0.86$). However, the high α–value of TFE ($\alpha = 1.51$) does not translate into efficient geminate recombination. We propose that any correlation between hydrogen-bonding and geminate recombination is compromised in TFE by the increased propensity for this solvent to exist as a monomer compared to water and ethanol.[47,48] This tendency arises from the ability of TFE to form an intramolecular hydrogen bond between the hydroxyl proton and the electrophilic CF$_3$-group.[49] The preference of TFE to exist as a monomer is evident in IR absorption spectra. Figure 3 presents static IR-absorption spectra of the OH stretching region for 320 mM solutions of ethanol and TFE dissolved in CCl$_4$. The assignments of the transitions observed in this region of the spectrum are in accord with previous studies.[48,50] In ethanol, the transition located at 3631 cm^{-1} is assigned as the 'monomer' band arising from hydroxyl groups not involved in hydrogen bonding. The transition located at 3493 cm^{-1} is assigned to hydroxy groups involved in intermolecular hydrogen-bonds resulting in dimer formation. Finally, the broad band centered at 3336 cm^{-1} is assigned to polymeric structures where the hydroxyl group serves as both a hydrogen-bonding donor and acceptor. Similar assignments have been made for TFE.[51] Inspection of the intensities in ethanol and TFE demonstrates that the extent of polymer formation is reduced in TFE.[52] The tendancy for TFE to self associate is also reflected by the modest hydrogen-bond accepting ability of this solvent, a behavior that would further frustrating intermolecular bonding.[46] Therefore, the geminate-recombination quantum yields reported here suggest that the efficiency of recombination is greatly influenced by the propensy of the solvent to self-associate.

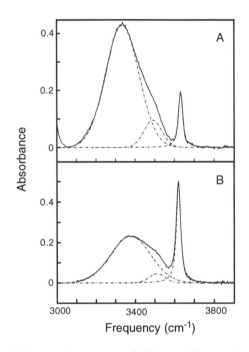

Figure 3. Infrared absorption spectra of OH stretching region for 340 mM solutions of ethanol (A) and 2,2,2-trifluoroethanol (B) dissolved in CCl₄ at 293 K. Dashed linse represent fits in which the intensity in this frequency region is decomposed in to three transitions representing monomer, dimer, and polymer bands as discussed in the text.

It should be noted that the TRRR studies in TFE indicate that the dynamics observed in this solvent may be unique such that comparison with other solvents must be performed with caution. The delayed recombination in TFE combined with the observation of ClO scattering in this solvent suggests that following photodissociation, the photofragments are stabilized through solvent/photofragment complexation. Furthermore, the depletion and recovery in TFE scattering intensity at 830 cm^{-1} indicates that it may be more appropriate to envision photoexcitation of a OClO-TFE complex rather than weakly solvated

OClO. Delayed geminate recombination due to solvent complexation has been observed during the formation of A'-state I_2.[12] Solvent complexation of ClO has been observed previously in benzene demonstrating that complexation in TFE is not unreasonable, and may be responsible for the reduced recombination in this solvent.[53]

Vibrational Relaxation

The results presented here, combined with those of previous studies, demonstrate that the rate constant for OClO vibrational relaxation is solvent dependent. Previous time-resolved resonance Raman anti-Stokes studies found that OClO intermolecular vibrational relaxation occurs in ~9 ps in water and ~33 ps in acetonitrile.[42] The increased vibrational relaxation rate in water relative to acetonitrile was suggested to arise from solvent-solute hydrogen bonding. Here, the vibrational-relaxation time in ethanol is ~30 ps, similar to the timescale observed in acetonitrile.   The similarity in relaxation time constants between ethanol and acetonitrile apparently invalidates simple correlation between vibrational relaxation and the presence of intermolecular hydrogen bonds. However, the symmetric stretch (918 cm^{-1}) and methyl-rock (1124 cm^{-1}) modes of acetonitrile are well matched to the OClO symmetric (945 cm^{-1}) and asymmetric stretch (1100 cm^{-1}) modes such that the coupling between OClO and acetonitrile is expected to be appreciable. Therefore, the ~30-ps relaxation time observed in acetonitrile may represent an upper limit for polar, aprotic solvents. In TFE, the ~80-ps vibrational relaxation time constant is the longest relaxation time for solution-phase OClO observed to date. We propose that the protracted vibrational relaxation time in TFE reflects the inefficiency of intermolecular hydrogen bonding in this solvent. As mentioned above, TFE undergoes a high degree of self-association such that the extent in intermolecular hydrogen bonding is less than expected from the Kalmut-Taft α-value. In addition, the OClO symmetric-stretch (ν_1) frequency is sensitive to intermolecular hydrogen bonding.   For example, the symmetric-stretch fundamental transition is observed at 945 cm^{-1} in water, but decreases to 938 cm^{-1} in acetonitrile and cyclohexane.[42,45,54] The frequency of this transition is 940 cm^{-1} in both ethanol and TFE, consistent with modest intermolecular hydrogen bonding in these solvents.

Acknowledgments

We thank Paul Blainey for acquiring the data presented in Figure 3. The National Science Foundation is acknowledged for their support of this work through the CAREER program (CHE-9701717 and CHE-0091320). Acknowledgement is made to the donors of the Petroleum Research Fund, administered by the American Chemical Society. PJR is a Cottrell Fellow of the Research Corporation and an Alfred P. Sloan Fellow. CLT gratefully acknowledges the University of Åarhus for financial support.

References

1. Oxtoby, D. W. *Adv. Chem. Phys.* **1981**, *47*, 487.
2. Owrutsky, J. C.; Raftery, D.; Hochstrasser, R. M. *Annu. Rev. Phys. Chem.* **1994**, *45*, 519.
3. Laubereau, A.; Kaiser, W. *Rev. Mod. Phys.* **1978**, *50*, 607.
4. Paige, M. E.; Harris, C. B. *Chem. Phys.* **1990**, *149*, 37.
5. Harris, A. L.; Brown, J. K.; Harris, C. B. *Annu. Rev. Phys. Chem.* **1988**, *39*, 341.
6. Chuang, T. J.; Hoffman, G. W.; Eisenthal, K. B. *Chem. Phys. Lett.* **1974**, *25*, 201.
7. Bado, P.; Dupuy, C.; Magde, D.; Wilson, K. R.; Malley, M. M. *J. Chem. Phys.* **1984**, *80*, 5531.
8. Zadoyan, R.; Li, Z.; Ashjian, P.; Martens, C. C.; Apkarian, V. A. *Chem. Phys. Lett.* **1994**, *218*, 504.
9. Yan, Y.-J.; Whitnell, R. M.; Wilson, K. R.; Zewail, A. H. *Chem. Phys. Lett.* **1992**, *193*, 402.
10. Nesbitt, D. J.; Hynes, J. T. *J. Chem. Phys.* **1982**, *77*, 2130.
11. Xu, X.; Yu, S.-C.; Lingle, R.; Zhu, H.; Hopkins, J. B. *J. Chem. Phys.* **1991**, *95*, 2445.
12. Lingle, R. J.; Xu, X.; Yu, S.-C.; Zhu, H.; Hopkins, J. B. *J. Chem. Phys.* **1990**, *93*, 5667.
13. Scherer, N. F.; Jonas, D. M.; Fleming, G. R. *J. Chem. Phys.* **1993**, *99*, 153.
14. Otto, B.; Schroeder, J.; Troe, J. *J. Chem. Phys.* **1984**, *81*, 202.
15. Batista, V. S.; Coker, D. F. *J. Chem. Phys.* **1996**, *105*, 4033.
16. Schwartz, B. J.; King, J. C.; Zhang, J. Z.; Harris, C. B. *Chem. Phys. Lett.* **1993**, *203*, 503.
17. Kliner, D. A. V.; Alfano, J. C.; Barbara, P. F. *J. Chem. Phys.* **1993**, *98*, 5375.
18. Walhout, P. K.; Alfano, J. C.; Thakur, K. A. M.; Barbara, P. F. *J. Phys. Chem.* **1995**, *99*, 7568.
19. Benjamin, I.; Whitnell, R. M. *Chem. Phys Lett.* **1993**, *204*, 45.
20. Benjamin, I.; Barbara, P. F.; Gertner, B. J.; Hynes, J. T. *J. Phys. Chem.* **1995**, *99*, 7557.
21. Khune, T.; Vohringer, P. *J. Chem. Phys.* **1996**, *105*, 10788.
22. Banin, U.; Waldman, A.; Rhuman, S. *J. Chem. Phys.* **1992**, *96*, 2416.
23. Banin, U.; Rhuman, S. *J. Chem. Phys.* **1993**, *98*, 4391.
24. Hess, S.; Bursing, H.; Vohringer, P. *J. Chem. Phys.* **1999**, *111*, 5461.
25. Hamm, P.; Lim, M.; Hochstrasser, R. M. *J. Chem. Phys.* **1997**, *107*, 10523.

26. Heilweil, E. J.; Doany, F. E.; Moore, R.; Hochstrasser, R. M. *J. Chem. Phys.* **1982**, *76*, 5632.
27. Pugliano, N.; Szarka, A. Z.; Gnanakaran, S.; Triechel, M.; Hochstrasser, R. M. *J. Chem. Phys.* **1995**, *103*, 6498.
28. Gnanakaran, S.; Hochstrasser, R. M. *J. Chem. Phys.* **1996**, *105*, 3486.
29. Whitnell, R. M.; Wilson, K. R.; Hynes, J. T. *J. Chem. Phys.* **1992**, *96*, 5354.
30. Walhout, P. K.; Silva, C.; Barbara, P. F. *J. Phys. Chem.* **1996**, *100*, 5188.
31. Li, M.; Owrutsky, J.; Sarisky, M.; Culver, J. P.; Yodh, A.; Hochstrasser, R. M. *J. Chem. Phys.* **1993**, *98*, 5499.
32. Thomsen, C. L.; Madsen, D.; Thogersen, J.; Byberg, J. R.; Keiding, S. R. *J. Chem. Phys.* **1999**, *111*, 703.
33. Coon, J. B. *J. Chem. Phys.* **1946**, *14*, 665.
34. Vaida, V.; Simon, J. D. *Science* **1995**, *268*, 1443.
35. Thomsen, C. L.; Reid, P. J.; Keiding, S. R. *J. Am. Chem. Soc.*, submitted,
36. Thogersen, J.; Thomsen, C. L.; J. Aa. Poulsen; Keiding, S. R. *J. Phys. Chem. A* **1998**, *102*, 4186.
37. Poulsen, J. A.; Thomsen, C. L.; Keiding, S. R.; Thogersen, J. *J. Chem. Phys.* **1998**, *108*, 8461.
38. Thorgersen, J.; Jepsen, P. U.; Thomsen, C. L.; Poulsen, J. A.; Byberg, J. R.; Keiding, S. R. *J. Phys. Chem.* **1997**, *101*, 3317.
39. Thomsen, C. L.; Philpott, M. P.; Hayes, S. C.; Reid, P. J. *J. Chem. Phys.* **2000**, *112*, 505.
40. Philpott, M. J.; Charalambous, S.; Reid, P. J. *Chem. Phys.* **1998**, *236*, 207.
41. Philpott, M. J.; Charalambous, S.; Reid, P. J. *Chem. Phys. Lett.* **1997**, *281*, 1.
42. Hayes, S. C.; Philpott, M. P.; Mayer, S. G.; Reid, P. J. *J. Phys. Chem. A* **1999**, *103*, 5534.
43. Hayes, S. C.; Philpott, M. J.; Reid, P. J. *J. Chem. Phys.* **1998**, *109*, 2596.
44. Chang, Y. J.; Simon, J. D. *J. Phys. Chem.* **1996**, *100*, 6406.
45. Esposito, A.; Foster, C.; Beckman, R.; Reid, P. J. *J. Phys. Chem. A* **1997**, *101*, 5309.
46. Kamlet, M. J.; Abboud, J.-L. M.; Abraham, M. H.; Taft, R. W. *J. Org. Chem.* **1983**, *48*, 2877.
47. Kivinen, A.; Murto, J. *Suomen Kemistilehti B* **1967**, *40*, 6.
48. Laenen, R.; Rauscher, C. *J. Chem. Phys.* **1997**, *107*, 9759.
49. Radnai, T.; Ishigoro, S.; Ohtaki, H. *J. Solut. Chem.* **1989**, *18*, 771.
50. Sutherland, G. B. B. M. *Faraday Discuss. Chem. Soc.* **1950**, *9*, 889.
51. Perttila, M. *Spectrochimica Acta* **1979**, *35A*, 585.
52. The extinction coefficient of the monomer band in TFE is greater than in ethanol; therefore, care must be taken in any direct comparison between ethanol and TFE. However, comparison of the relative monomer to polymer band intensities demonstrates that the extent of polymer formation in ethanol is greater relative to TFE.
53. Dakhnovskii, Y. I.; Doolen, R.; Simon, J. D. *J. Chem. Phys.* **1994**, *101*, 6640.
54. Foster, C. E.; Reid, P. J. *J. Phys. Chem. A* **1998**, *102*, 3541.

Chapter 11

Solvent Dependence of Excited State Lifetimes in 7-Dehydrocholesterol and Simple Polyenes

Neil A. Anderson and Roseanne J. Sension[*]

Department of Chemistry, University of Michigan, 930 North University, Ann Arbor, MI 48109–1055

Ultrafast transient absorption spectroscopy has been used to study the internal conversion dynamics of 7-dehydrocholesterol (provitamin D_3) as a function of solvent. Solvent polarity and viscosity have only a small effect on the observed dynamics. These results are compared with measurements on simple polyene analogs of 7-dehydrocholesterol and previtamin D_3.

Photochemical reactions of polyene chromophores play a key role in the function of many important biological systems. Prototypical examples include the *cis-trans* isomerization reactions of retinal chromophores and the electrocyclic ring-opening reaction of 7-dehydrocholesterol (provitamin D_3). In accord with the importance of such reactions, the photochemical reactions of simple analogs, such as 1,3,5-hexatriene and 1,3-cyclohexadiene have been studied extensively in recent years, both by our laboratory (*1-5*) and by others (*6-15*). A key feature observed for these small polyenes is ultrafast (<60 fs) internal conversion from the initially excited state (1^1B) to an optically forbidden state of lower energy (2^1A or S_1) followed by internal conversion to the ground electronic state (190 - 500 fs). Similar excited state dynamics have been reported following excitation of 7-dehydrocholesterol (DHC) leading to ring-opening and the formation of previtamin D_3 (*16-17*). Excitation of DHC dissolved in methanol results in the formation of an excited electronic state characterized by a broad absorption band in the visible region of the spectrum. This excited state absorption decays on a time scale of 950 fs, with the appearance of an ultraviolet difference spectrum characteristic of formation of previtamin D_3 from DHC on the same time scale (*17*). Vibrational relaxation of previtamin D_3 occurs on a 1-5 ps time scale, while conformational relaxation producing an equilibrium mixture of the all cis, cZc, and mono-s-trans, cZt, conformational isomers, as illustrated in Figure 1, occurs on a time scale of *ca.* 100 ps at 22°C.

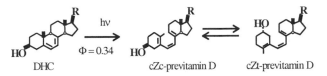

Figure 1. Photoinduced ring-opening reaction and subsequent ground state relaxation observed following excitation of DHC. (R = C₈H₁₇)

In the work reported here we extend our earlier investigations of DHC, and in particular the excited state absorption of DHC, to include both polar and non-polar solvents. A full analysis of the UV and visible transient absorption changes is reported for DHC dissolved in n-heptane. The visible transient absorption signal characteristic of population in the 2A state was also studied in 2-butanol and n-hexadecane. The results are compared with measurements on simple cyclo-hexadiene and hexatriene polyene analogs of DHC and pre-vitamin D_3.

Experimental

The photoinitiated reaction of 7-dehydrocholesterol to form previtamin D_3 was studied using pump-probe transient absorption spectroscopy. The output of a home-built 1 kHz regeneratively amplified titanium sapphire laser was split into two to provide pump and probe beams. For all scans, the third harmonic of the laser at *ca.* 265 nm was generated as the pump beam. Pump energy at the sample was 300-600 nJ/pulse. Experiments were typically carried out with a 150 fs FWHM pulse at the sample. Visible probe wavelengths were obtained from the output of a home-built non-collinear optical parametric amplifier (OPA) con-structed after the design of Riedle and coworkers (*18*). Ultraviolet probe wave-lengths were generated by frequency doubling the output of the OPA in a β-barium borate (BBO) nonlinear optical crystal. Except when specified for anisotropy measurements, all scans were obtained with the relative polarizations of the pump and probe beams at the magic angle of 54.7° to eliminate contributions due to changes in the orientation of the probed transition dipole. In all cases, the probe beam was attenuated to be <10% of the pump beam intensity. The data were collected using digital noise reduction procedures described previously *(19)*.

7-Dehydrocholesterol (Aldrich) was used as received and dissolved in spectroscopic or HPLC grade solvents (methanol, 2-butanol, heptane or hexadecane) at a concentration of approximately 10^{-3} M. Experiments were carried out at 295 K. The sample was flowed through a 1 mm path length cell to refresh the volume between laser shots. Static absorption spectra of samples were taken before and after use to assure that the samples had not degraded substantially through a buildup of the previtamin photoproduct.

Results

Transient absorption traces were obtained for DHC dissolved in n-heptane for 13 probe wavelengths between 265 nm and 340 nm for time delays extending to 300 ps. Typical data are shown and summarized in Figure 2. The heptane data are qualitatively similar to the methanol data reported previously and consequently the data were analyzed in the same manner *(17)*.

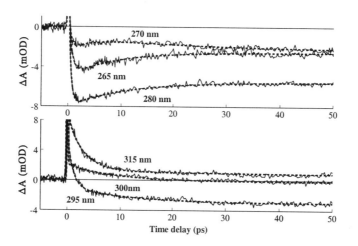

Figure 2. Typical transient absorption signals obtained following excitation of DHC dissolved in n-heptane using the specified probe wavelengths. The dashed lines represent the fit of the data to a model consisting of a an instrument-limited solvent absorption, three exponentially decaying components and a permanent absorption change at long times.

Global fitting of the data to a sum of exponentials requires a subpicosecond component, a wavelength dependent 2-9 ps component, a >50 ps decay component, and a non-decaying component. The subpicosecond component observed in the UV transient absorption signal represents an excited state absorption signal, decaying on a 600 ± 200 fs time scale.

The intermediate picosecond component is dominant at all probe wavelengths. This component has a wavelength dependent time constant. Within the DHC absorption band, the time constant is 6-9 ps. At longer wavelengths, it is distinctly faster, with time constants of 2.6 ps at 308 nm, 3.2 ps at 315 nm, and 2.0 ps at 340 nm obtained from the least-squares analysis. This component corresponds to the vibrational relaxation or cooling component as described for DHC in methanol *(17)*. The rate of cooling may be slightly slower in heptane than in methanol, where observed time constants range from just over 1 ps to 5 ps.

An additional longer time scale (>50 ps) relaxation component is also observed in the data. This component corresponds to a blue shift of the photoproduct spectrum. The spectral signature of this component is similar to the 100 ps spectral shift observed in methanol or ethanol *(16,17)*, although the amplitude is substantially smaller. As was observed in methanol, the difference spectrum at 300 ps is consistent with the formation of pre-vitamin D_3 from DHC. This is illustrated in Figure 3.

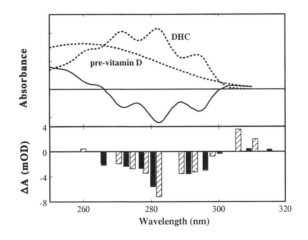

Figure 3. Top panel: The steady state absorption spectra observed for DHC and pre-vitamin D_3 along with the anticipated steady state difference spectrum. Bottom panel: A bar plot illustrating the observed absorption differences at 300 ps in methanol (cross hatched bars) and heptane (solid bars).

Visible Excited State Absorption

The S_1 to S_n absorption of 7-dehydrocholesterol in the near UV and visible provides a simple determination of the S_1 (2^1A) excited state lifetime. This absorption signal was initially identified and assigned in our earlier work on DHC dissolved in methanol *(17)*. The visible excited state absorption (ESA) signal is not complicated by the presence of the ground state cooling and rotational isomerization components present in the UV data. In the present paper we report measurements of this excited state absorption in heptane, hexadecane, and 2-butanol. These measurements were performed to investigate the effects of solvent viscosity and polarity on the excited state lifetime. Data in each of these solvents were collected at probe wavelengths of 500 and 540 nm. Transient absorption traces obtained at 500 nm are shown in Figure 4.

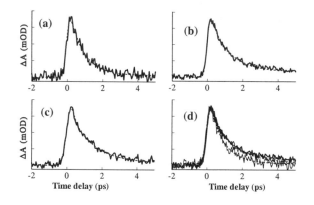

Figure 4. Transient absorption data obtained at 500 nm following the excitation of DHC at 268 nm. (a) DHC in heptane, (b) hexadecane, (c) 2-butanol. Normalized data in all three solvents are compared in panel (d). The absorption signal decays substantially faster in n-heptane than in the other two solvents studied here.

These two wavelengths were chosen to minimize the contribution to the signal arising from two photon absorption of the solvent, while maximizing the magnitude of the observed ESA signal. Excitation wavelengths around 268 nm are sufficiently energetic to lead to direct two-photon excitation of the solvent, either visible + UV or UV + UV. Care must be exercised to limit two photon excitation by the pump. In addition the visible + UV absorption of the solvent must be characterized. Solvent-only scans of heptane and hexadecane showed no instrument limited spike at 500 nm or 540 nm. In 2-butanol, a very small solvent spike was observed at 500 nm, but not at 540 nm. All three solvents exhibit a slight instrument limited increase in absorption attributed to two photon excitation of the solvent. As in methanol *(17)* no decay of the solvent-only absorption signal was observed over the time window collected. The magnitude of the absorption was again comparable with the size of the long-lived plateau in the DHC scans.

All scans are modeled well using a single exponential decay of the absorption and a long-lived plateau. The slight solvent spike observed in 2-butanol at 500 nm is also included in the fit. The time constants obtained from the data are summarized in Table I.

Table I. ESA Decay Parameters as a Function of Solvent.

Solvent	Viscosity[20] (mPa s) 20°C	τ (500 nm) (ps)	τ (540 nm) (ps)	anisotropy r(t=0)
Methanol[a]	0.55	0.95 ± 0.10		0.21(600 nm)
2-butanol	4.21	1.28	1.17	0.21 (540 nm) 0.29 (500 nm)
heptane	0.42	0.80	0.81	0.28 (500 nm)
hexadecane	3.34	1.03	1.08	0.30 (500 nm)

a. Data in methanol were reported previously *(17)*. The time constant was obtained from a global analysis of 6 probe wavelengths between 470 nm and 650 nm. The anisotropy was determined only at 600 nm.

The data in Table I show a clear, although small, solvent effect on the S_1 lifetime for DHC. The S_1 lifetime ranges from 0.8 ps to 1.2 ps over the range of solvent viscosity and polarity studied. Solvent viscosity appears to influence the lifetime, although an order of magnitude increase in viscosity results only in a 20-25% increase in the S_1 lifetime. A small viscosity effect is consistent with the prediction that accessing the S_1-S_0 conical intersection does not require any large scale motion which would be greatly hindered by a viscous solvent *(21)*. Any effect of solvent polarity on the excited state lifetime must be even smaller. The time constants observed for the decay of the S_1 state of DHC in alcohols and alkanes of similar viscosity are similar.

In addition to the magic angle transient absorption signals described above, the absorption anisotropy was measured for DHC in each solvent using a 500 nm probe pulse. The anisotropy was also measured at 540 nm for DHC in 2-butanol. In all cases, the anisotropy remained relatively constant over at least the first 2 ps. Decay of the anisotropy is observed after this, as the slight absorption plateau begins to dominate the signal. This observation is once again consistent with the data obtained in methanol *(17)*. The average values of the anisotropy over the first 1 ps are reported in the last column of Table I. The anisotropy values reported in Table I are solvent independent, but wavelength dependent. This observation suggests the presence of two or more overlapping S_1 to S_n transitions contributing to the visible signal.

Discussion

The visible and ultraviolet transient absorption data obtained for DHC in the nonpolar solvent n-heptane are qualitatively identical to the transient absorption data obtained in methanol *(17)*. Although the precise time constants differ quantitatively, even these differences are small. The overall similarity of the UV data in the two solvents indicates that a similar model should apply. Time scales vary slightly, but the same relaxation processes are observed. The polarity of the solvent has no significant effect on the photochemistry.

The fastest observed time constants are attributed to internal conversion from the S_1 state to the ground state. In heptane the time constants are 0.80 ps, visible, 0.6 ps UV, in methanol these time constants are slightly longer: 0.95 ps, visible, 0.7 ps UV. The visible and UV time constants in each solvent agree within the uncertainty of the UV measurement, where the data is complicated by an instrument-limited solvent spike and vibrational relaxation of the solute. The lifetime of the 2^1A (S_1) state in DHC is substantially longer than the lifetime for the isolated cyclohexadiene or hexatriene chromophores *(4,5)*. The influence of solvent on internal conversion will be discussed in greater detail below.

The wavelength dependent picosecond component is assigned to vibrational relaxation of hot DHC and pre-vitamin D_3 on the ground state surface following internal conversion. This component is 2-9 ps in heptane and somewhat faster (1-5 ps) in methanol. The evolution of the time constant, becoming faster at longer probe wavelengths, along with the sign of the signal, corresponding to an absorption increase below 290 nm and an absorption decrease above 300 nm, allow assignment of this component to vibrational energy redistribution and thermal-ization, although it is likely that some rotational isomerization of the photoproduct also occurs while the molecule is vibrationally excited.

Finally, a small amplitude component of > 50 ps in heptane and 100 ± 20 ps in methanol was observed. Our analysis of this component corroborates a previous assertion that it results from the barrier crossing for rotational isomerization in the product previtamin as the system attains a thermal distribution of rotamers *(16)*.

The wavelength dependent sign of this component demonstrates that it corresponds to a blue-shift of the spectrum. The observation of a blue-shifting spectrum is consistent with the expectation that the initially formed cZc conformer of previtamin D_3 will absorb at longer wavelengths than the cZt conformer. The >50 ps evolution of the spectrum corresponds to the formation of a thermal equilibrium mixture of the cZc- and cZt conformers of previtamin D_3 from the initially prepared cZc conformer. The small amplitude of this equilibration in heptane suggests that more cZt conformer is formed in the initial vibrational relaxation stage in heptane than in methanol. This is consistent with the relative time scales for vibrational relaxation in the two solvents, *ca.* 9 ps in heptane and *ca.* 5 ps in methanol.

Influence of solvent on excited state lifetime

The fundamental questions being addressed in our ongoing study of ground and excited state dynamics in polyene chromophores are designed to explore the effect of the internal energy bath (vibrational degrees of freedom of the molecule itself) and of coupling to an external solvent bath on the observed dynamics. Some general conclusions concerning influence of solvent on excited state lifetime may now be drawn from our studies of cyclohexadiene *(1,3,4)*, hexatriene *(2,4,5)* and DHC *(17)*. Results from our investigations and the investigations of other groups are summarized in Table II. All of the polyenes studied have similar excited state lifetimes. The optically accessible 1^1B state has a very short lifetime (<<100 fs). This suggests that motion along the initial Franck-Condon active vibrational coordinates leads to internal conversion from the optically accessible 1^1B state to the lower energy 2^1A state. No influence of environment on the rate of this internal conversion process is apparent, although future experiments with better time resolution may reveal modest solvent and chromophore differences.

The 2^1A (S_1) state has a lifetime on the order of a picosecond (0.2 - 1.2 ps) in all of the molecules and solvents summarized in Table II. Conical intersections facilitating rapid internal conversion between the 2^1A excited state and the ground electronic state have been calculated for each of these molecules. The rate for internal conversion is controlled by accessibility of the conical intersection and by vibrational energy redistribution in the electronically excited molecule. With the exception of *cis*-hexatriene, the excited state lifetime in each of the systems investigated is largely independent of environment, with variations of ± 20%. These observations suggest that the path to the conical intersection is effectively barrierless and involves only small amplitude motions of the solute.

Acknowledgments

This research is supported by the NSF (CHE-0078972). NAA was supported by the Center for Ultrafast Optical Science under grant STC-PHY-8920108.

Table II. Summary of Excited State Lifetimes in Hexatriene, Cyclohexadiene and 7-dehydrocholesterol.

Molecule	Phase or Solvent	1B lifetime (fs)	2A lifetime (fs)	Conical Intersection Calculated
cyclohexadiene	cyclohexane	$\geq 10^{(22)}$	$\sim 300^{(4)}$	(23)
	hexadecane		$<300^{(4)}$	
	ethanol		$<300^{(10)}$	
	vapor	$60^{(8)}$	$\sim 200^{(8)}$	
DHC	methanol	$<<100^{(17)}$	$950^{(17)}$	(21)
	2-butanol	$<<100$	1200	
	heptane	$<<100$	800	
	hexadecane	$<<100$	1050	
cis-hexatriene	cyclohexane	$20^{(24)}$-$50^{(5)}$	$250^{(5)}$	(25)
	hexadecane		$<250^{(4)}$	
	ethanol		$470^{(10)}$	
	vapor	$15^{(24)}$	$730^{(7)}$	
trans-hexatriene	cyclohexane	$20^{(26)}$-$55^{(5)}$	$190^{(5)}$	(23)
	acetonitrile		$<500^{(9)}$	
	vapor	$40^{(27)}$	$270^{(7)}$	

References

1. Pullen, S.; Walker, L.A., II; Donovan, B.; Sension, R.J. *Chem. Phys. Lett.* **1995**, *242*, 415-420.
2. Pullen, S.H.; Anderson, N.A.; Walker, L.A., II; Sension, R.J. *J. Chem. Phys.* **1997**, *107*, 4985-4993.
3. Pullen, S.H.; Anderson, N.A.; Walker, L.A., II; Sension, R.J. *J. Chem. Phys.* **1998**, *108*, 556-563.
4. Anderson, N.A.; Pullen, S.H.; Walker, L.A., II; Shiang, J.J.; Sension, R.J. *J. Phys. Chem. A* **1998**, *102*, 10588-10598.
5. Anderson, N.A.; Durfee, C.G., III; Murnane, M.M.; Kapteyn, H.C.; Sension, R.J. *Chem. Phys. Lett.* **2000**, *323*, 365-371.
6. Hayden, C.C.; Chandler, D.W. *J. Phys. Chem.* **1995**, *99*, 7897-7903.
7. Cyr, D.R.; Hayden, C.C; *J. Chem. Phys.* **1996**, *104*, 771-774.
8. Fuss, W.; Schikarski, T.; Schmid, W.E.; Trushin, S.A.; Kompa, K.L. *Chem. Phys. Lett.* **1996**, *262*, 675-682.
9. Ohta, K.; Naitoh, Y.; Saitow, K.; Tominaga, K.; Hirota, N.; Yoshihara, K. *Chem. Phys. Lett.* **1996**, *256*, 629-634.
10. Lochbrunner, S.; Fuss, W.; Kompa, K.L.; Schmid, W.E.. *Chem. Phys. Lett.* **1997**, *274*, 491-498.
11. Fuss, W.; Schikarski, T.; Schmid, W.E.; Trushin, S.; Hering, P.; Kompa, K.L. *J. Chem. Phys.* **1997**, *106*, 2205-2211.
12. Fuss, W.; Lochbrunner, S. *J. Photochem. Photobiol. A* **1997**, *105*, 159-164.
13. Trushin, S.A.; Fuss, W.; Schikarski, T.; Schmid, W.E.; Kompa, K.L. *J. Chem. Phys.* **1997**, *106*, 9386-9389.
14. Lochbrunner, S.; Fuss, W.; Schmid, W.E.; Kompa, K.L. *J. Phys. Chem. A* **1998**, *102*, 9334-9344.
15. Ohta, K.; Naitoh, Y.; Tominaga, K.; Hirota, N.; Yoshihara, K. *J. Phys. Chem. A* **1998**, *102*, 35-44.
16. Fuss, W.; Höfer, T.; Hering, P.; Kompa, K.L.; Lochbrunner, S.; Schikarski, T.; Schmid, W.E. *J. Phys. Chem.* **1996**, *100*, 921-927.
17. Anderson, N.A.; Shiang, J.J.; Sension, R.J. *J. Phys. Chem. A* **1999**, *103*, 10730-10736.
18. Wilhelm, T.; Piel, J.; Riedle, E. *Opt. Lett.* **1997**, *22*, 1494-1496.
19. Anderson, N. A., Ph.D. Thesis, University of Michigan, Ann Arbor, MI, **2000**.
20. Lide, D., Ed. *Handbook of Chemistry and Physics 71st Edition*, (CRC Press, Boca Raton, 1991).
21. Bernardi, F.; Olivucci, M.; Ragazos, I. N.; Robb, M. A. *J. Am. Chem. Soc.* **1992**, *114*, 8211.

22. Trulson, M. O; Dollinger, G. D.; Mathies, R. A. *J. Chem. Phys.* **1989**, *90*, 4274-4281.
23. Garavelli, M.; Bernardi, F.; Olivucci, M.; Vreven, T.; Klein, S.; Celani, P.; Robb, M. A. *Faraday Discussions*, **1998**, *110*, 51.
24. Ci, X.; Myers, A.B. *J. Chem. Phys.* **1992**, *96*, 6433-6442.
25. Olivucci, M.; Bernardi, F.; Celani, P.; Ragazos, I.; Robb, M. A. *J. Am. Chem. Soc.* **1994**, *116*, 1077-1085.
26. Ci, X.; Pereira, M. A.; Myers, A. B., *J. Chem. Phys.* **1990**, *92*, 4708-4717.
27. Myers, A. B.; Pranata, K. S., *J. Phys. Chem.* **1989**, *93*, 5079-5087.

Vibrational Dynamics

Chapter 12

Vibrational Dynamics in Porous Silica Glasses Studied by Time-Resolved Coherent Anti-Stokes Raman Scattering

Keisuke Tominaga[1,*], Hiroaki Okuno[1], Hiroaki Maekawa[1], Tadashi Tomonaga[1], Brian J. Loughnane[2], Alessandra Scodinu[2], and John T. Fourkas[2,*]

[1]Department of Chemistry, Faculty of Science, Kobe University, Nada, Kobe 657–8501, Japan
[2]Eugene F. Merkert Chemistry Center, Boston College, 140 Commonwealth Avenue, Chestnut Hill, MA 02467

We have studied pore size dependence on the vibrational dephasing of the CD stretching of $CDCl_3$ in porous silica glasses by time-resolved coherent anti-Stokes Raman scattering (CARS) technique. The CARS signal decays almost single-exponentially for all the porous silica glasses, and the decay time constant becomes smaller by decreasing the pore size. There is almost no spectral shift in the IR spectra of the CH stretching of $CHCl_3$ when the liquid is confined in the glasses. We explain these observations in terms of a two-state model in which liquid molecules inside the pore are classified into "surface molecules" and "bulk molecules".

Porous silica glasses are materials with nano-scale pores that have several characteristic features; the diameters of the pores range from a few tens of angstroms to several thousands of angstroms, depending on its preparation method; the pore sizes can be controlled and are quite uniform; many kinds of liquid molecules can go into the glass and be trapped in the pores. Because of these characteristic features, porous silica glasses are a suitable system to investigate the confinement effect on molecular motions. So far, confinement effects on molecular dynamics in porous silica glasses have been studied by means of NMR (*1-4*), Raman scattering (*5-8*), and femtosecond optical Kerr effect (*9-12*).

Local structures and dynamics in liquids are often studied by vibrational spectroscopy. This is because vibrational spectra are very sensitive to the environment around the oscillator. The transition frequency and lineshape of the spectrum, or vibrational dephasing, are good "detectors" to monitor microscopic details in liquids. In this work we study vibrational dynamics of liquids confined in porous silica glasses by time-resolved CARS measurement and FT-IR measurements.

Experimental

The preparation of the porous silica glasses was reported elsewhere in detail (*10*). The porous silica glasses were heated at 450 °C for a few hours, cooled down in the desiccater, and filled with the solvent. The sample cell was sealed with an epoxy glue.

The CARS experiment was performed with a pair of synchronously pumped dye lasers. A schematic picture of the system is shown in Figure 1. One oscillator gave pulses with a duration of about 90 fs and a center wavelength of 600 nm. The duration of the other oscillator was set to be about 8 ps to avoid a timing jitter between the two lasers. The wavelength of the picosecond laser is tunable from 600 to 750 nm. The details of the oscillators (*13*) and the amplifier (*14*) for the femtosecond laser pulse were already mentioned elsewhere. The energy of the amplified femtosecond pulse is about 5 μJ/pulse. The picosecond pulse was amplified by a standard method using a three-stage dye amplifier. The green output from the Nd:YAG regenerative amplifier pumps both the amplifiers. The first and second stages consist of flow cells with a 1 mm optical path length and solutions of Pyridine1 in a mixture of ethanol and water. The third stage, which has a flow cell of a 10 mm optical path length and the same solution, is pumped from both the sides. The final output energy of the amplified pulse is about 10 μJ/pulse. The femtosecond pulse is split into two portions, one being a pump pulse, and the other a probe pulse. These two pulses are focused into the sample together with the picosecond pulse which works as a Stokes pulse. The input angles of the pulses satisfy the phase-matching condition. The CARS signal is collected by a photomultiplier and amplified by a lock-in amplifier (SRS830).

IR measurements were made with a Perkin-Elmer Model SPECTRUM 1000 FT-IR spectrometer.

162

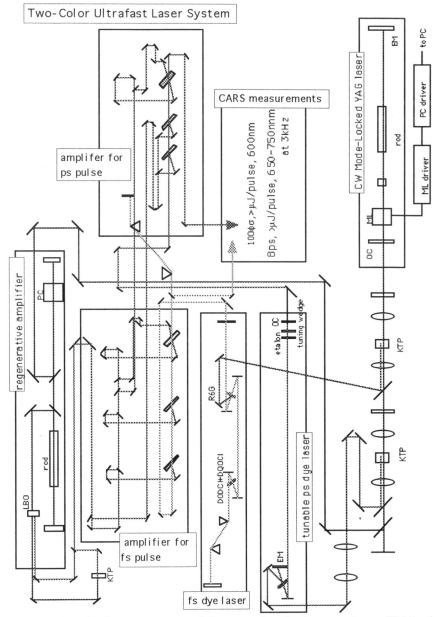

Figure 1. A schematic picture of the two-color dye laser system. EM (end mirror), OC (output coupler), ML (mode locker), and PC(Pockels cell).

Results

Time-Resolved CARS Measurements

Figure 2 shows a CARS signal from the CD stretching of $CDCl_3$ confined in a porous silica glass of a diameter of 24 Å. Since the polarizations of the pulses are under the magic angle condition, only the pure vibrational dephasing contributes to the signal. There is an instantaneous response due to an electronic polarizability at $t=0$, which is followed by a nuclear response at $t>400$ fs. The nuclear response at $0<t<400$ fs is obscured by the electronic response and cannot be discussed here. The nuclear response shows an exponential decay with a time constant of 0.89 ps, which is a half of the vibrational dephasing time. CARS signals from the liquids in other porous silica glasses and bulk liquid show a similar exponential behavior. Table I summarizes the obtained time constants of the CARS signals. As seen from the table, the time constant of the signal decay decreases by 5 to 15 % when the liquid molecules are confined in the porous silica glasses.

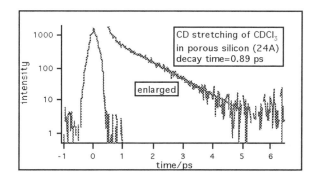

Figure 2.   The CARS signal from the CD stretching of $CDCl_3$ confined in a porous silica glass of a diameter of 24 Å

Table I.   Decay Time Constants of the CARS Signal from the CD Stretching of $CDCl_3$

pore diameter	decay time constant/ps
bulk liquid	1.05
59 Å	0.99
42 Å	0.93
24 Å	0.89

Steady-State IR Measurements

Figure 3 compares an IR spectrum of the C-H stretching of $CHCl_3$ in porous silica glass with that of the bulk solution. The solution is $CHCl_3$ dissolved in $CDCl_3$ (the concentration of $CHCl_3$ is 0.088 M). In Table II the peak frequency shift from that in the bulk liquid is shown for the solutions in five porous silica glasses. The peak frequency does not show a drastic change; it is slightly shifted to higher energies when the solution is trapped in the porous silica glass.

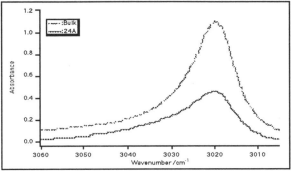

Figure 3.   IR spectra of the CH stretching of $CHCl_3$ in $CDCl_3$ (concentration = 0.88 M) of a bulk liquid and confined in a porous silica glass of a diameter of 24 Å

Table II.   Peak Shift and Spectral Width of the CH Stretching of $CHCl_3$ in $CDCl_3$ (concentration = 0.88 M) in Porous Silica Glasses

pore size/Å	$\delta v^a/cm^{-1}$	Δ^b/cm^{-1}
bulk	-	13.2
91	0.2±0.3	12.8
62	0.2±0.3	13.3
59	0.05±0.3	13.8
42	0.4±0.3	16.9
34	0.2±0.3	15.6
24	0.4±0.3	15.5

a: $\delta v = v_{peak} - v_{peak}$(bulk), where v_{peak} is the wavenumber of the peak position.
b: Δ is the full width at a half height of the spectrum.

Discussion

In the porous silica glasses decrease of the decay time constant of the CARS signal was observed. This is consistent with previous Raman studies on the

165

isotropic components of the C-H stretching of CHCl₃ (5), the ν_1 mode of CS₂ (6), and several vibrational bands of bezene-d₆ (7). Furthermore, dependence of the peak frequency of the IR spectrum on the pore size is found to be quite small. We explain these observations in terms of a two-state model in which the liquid molecules inside the pore are classified into "surface molecule" and "bulk molecule".

Two-State Model

Figure 4 shows a schematic picture to explain the two-state model. Liquid molecules in proximity to the surface behave differently from molecules inside because of the interaction with the surface. Therefore, the IR spectral shape can be expressed as a sum of two contributions. For simplicity we adopt a Lorentzian function for the spectral shape.

$$I(x) = a_1 \frac{\Delta_1}{(x-\delta/2)^2 + \Delta_1^2} + a_2 \frac{\Delta_2}{(x+\delta/2)^2 + \Delta_2^2} \tag{1}$$

Here, δ is the difference between the peak frequencies of the surface molecules and bulk molecules, and Δ_i and a_i correspond to the linewidth and the intensity of the i-th component.

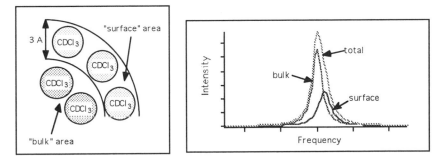

Figure 4. A schematic picture for the two-state model. left: the liquid molecules are classified into "surface molecules" and "bulk molecules". right: the observed spectrum is a sum of two contributions, one from the bulk molecules and the other from surface molecules.

It is well-known that the peak frequency and linewidth of the CD stretching of

CDCl$_3$ are affected by surrounding solvent molecules as shown in Table III (*15*). We tentatively calculate both the CARS signal and spectrum in terms of the two-state model with δ = 2 cm^{-1}, Δ_1 = 5.1 cm^{-1}, and Δ_2 = 10.2 cm^{-1}. We assume the thickness of the surface area is 3 Å, and a_1 and a_2 are determined by relative sizes of the surface and bulk areas. Δ_1 is equal to the linewidth of the bulk liquid, and Δ_2 is set be twice larger than Δ_1, implying that the molecular motion in proximity of the surface is slowed down due to some interaction with the surface such as a hydrogen bonding or mechanical hindrance. The purpose of the simulation is not to obtain these parameters precisely, but to explain the observations with physically reasonable values for the parameters.

Table III. Peak Frequencies and Isotropic Raman Line Widths (Full Width at a Half Height) of the CD Stretching Band of CDCl$_3$ in Solution (*15*)

solvent	*peak frequency/cm^{-1}*	*isotropic Raman line width/cm^{-1}*
CDCl$_3$ (neat)	2252	5.8±0.1
CH$_3$OH	2250	10.2±0.4
H$_2$O	2257	10.0±0.3
CH$_3$COCH$_3$	2253	8.7±0.1
C$_2$H$_5$OC$_2$H$_5$	2245	11.9±0.3
CCl$_4$	2251	6.1±0.1

Table IV shows results of the spectral simulation. Although the noticeable spectral broadening is observed in the porous silica glasses, the peak shift is fairly small. This is consistent with the observation. In Figure 5 the simulation of the CARS signal is shown. In the time region for the measurement ($0.4 < t < 5$ ps) the simulated signal decay can be approximated with a single exponential function. The obtained decay time constant is also shown in the figure. The trend of the decrease of the time constant is reproduced reasonably well though the simulations overestimate dependence of the decay time constant on the pore size compared to the observation. This simulation result suggests that the CH vibrational dynamics of chloroform is strongly affected by the surface probably due to a hard wall effect for molecules pointed at the wall. However, this surface effect on the vibrational dynamics is not "transferred" to the inside of the pore. This is probably because the vibrational dephasing of the CH stretching mode may be determined by a short-range collisional motion, and not by a long-range specific interaction such as hydrogen bonding between the chloroform molecules.

Table IV.   Results of the Spectral Simulation

pore size/Å	$\delta v^{a}/cm^{-1}$	$\Delta/\Delta (bulk)^{b}$
59	0.045	1.06
42	0.065	1.08
34	0.085	1.13
27	0.120	1.19
24	0.135	1.22

a: $\delta v = v_{peak} - v_{peak}(bulk)$, where v_{peak} is the wavenumber of the peak position.

b: Δ is the full width at a half height of the spectrum.

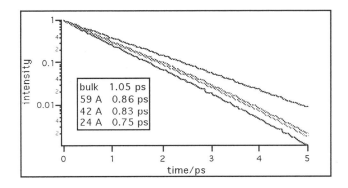

Figure 5.   Results of the simulation of the CARS signal by the two-state model. The obtained decay time constants are shown in the figure.

Finally we discuss the absolute intensity of the CARS signal from the porous silica glasses.   We estimate pore volumes relative to the whole glass volume to be about 50% from the absorbance of the CH stretching band.   This means that the about half of the space of the porous glasses is occupied by the pores.   The relative intensity of the CARS signal from the porous glass sample is about one-fourth of that from the bulk liquid.   This is consistent with the IR measurements because the CARS signal is proportional to the square of the concentration of the scattering medium.   This fact also confirms that the CARS signal results from the liquid molecules inside the pore but not from molecules near the surface due to the surface-enhanced Raman effect.

Acknowledgments

This work is supported by the U.S.-Japan Cooperative Science Program of JSPS and NSF.

References

1. Dore, J. C.; Dunn, M.; Hasebe, T.; Strange, J. H. *Colloids Surf.* **1989**, *36*, 199.
2. Liu, G.; Li, Y.; Jonas, J. *J. Chem. Phys.* **1989**, *90*, 5881.
3. Xu, S.; Zhang, J.; Jonas, J. *J. Chem. Phys.* **1992**, *97*, 4564.
4. Korb, J.-P.; Xu, S.; Cros, F.; Malier, L.; Jonas, J. *J. Chem. Phys.* **1997**, 107, 4044.
5. Nikiel, L.; Hopkins, B.; Zerda, T. W. *J. Phys. Chem.* **1990**, *94*, 7458.
6. Wallen, S. L.; Nikiel, L.; Yi, J.; Jonas, J. *J. Phys. Chem.* **1995**, *99*, 1542.
7. Yi, J.; Jonas, J. *J. Phys. Chem.* **1996**, *100*, 16789.
8. Czeslik, C.; Kim, Y. J.; Jonas, J. *J. Chem. Phys.* **1999**, *111*, 9739.
9. Farrer, R. A.; Loughnane, B. L.; Fourkas, J. T. *J. Phys. Chem. A* **1997**, 101, 4005.
10. Loughnane, B. L.; Farrer, R. A.; Fourkas, J. T. *J. Phys. Chem. B* **1998**, *102*, 5409.
12. Loughnane, B. L.; Farrer, R. A.; Scodinu, A.; Fourkas, J. T. *J. Chem. Phys.* **1999**, *111*, 5116.
13. Okamoto, H.; Yoshihara, K. *J. Opt. Soc. Am. B* **1990**, *7*, 1702.
14. Tominaga, K.; Keough, G. P.; Naitoh, Y.; Yoshihara, K. *J. Raman Spec.* **1995**, *26*, 495.
15. Tanabe, K.; Hiraishi, J. *Adv. in Molecular Relaxation and Interaction Processes* **1980**, *16*, 281.

Chapter 13

Third-Order Nonlinear Spectroscopy
of Coupled Vibrations

O. Golonzka, M. Khalil, N. Demirdöven, and A. Tokmakoff

Department of Chemistry, Massachusetts Institute of Technology, 77 Massachusetts Avenue, Cambridge, MA 02139

Frequency dispersed pump-probe and two-dimensional infrared photon echo spectroscopy have been used to investigate the inter-mode interactions in a model system of two coupled molecular vibrations. The amplitude of the quantum beats, observed in polarization sensitive pump-probe experiments, was used to determine the projection angle between the interacting dipoles. The frequency splitting between peaks in the two-dimensional photon echo spectrum is shown to be directly related to the strength of anharmonic couplings.

Introduction

Recent development of time-domain optical spectroscopic techniques is driven in-part by the possibility to follow in real time the structural changes occurring in the reactants during chemical processes (1). Structural information at each step of a reaction can be obtained by measuring the strength of interactions between different structural groups and identifying the mechanisms of these interactions. Vibrational spectroscopy can reveal such information by characterizing the nuclear potential governing the well-defined molecular motions and by examining the dynamics of these motions. Furthermore, nonlinear vibrational spectroscopic techniques, based on multiple interactions of the external fields with molecular transition dipoles, can directly examine the correlation between different vibrational motions (2-5) and yield information about the relative orientation of interacting structural groups (1, 6).

In this work we examine the processes contributing to the signal in third-order time resolved spectroscopies of coupled molecular vibrations and discuss how polarization sensitive third-order spectroscopy can be used to determine the projection angle between the transition dipole moments of coupled vibrations. We further demonstrate the effectiveness of 2-D photon echo spectroscopy in determining the strength of the coupling between different vibrations as encoded in the anharmonicities of the nuclear potential.

Several mechanisms can be identified as potential contributors to the nonlinear signals: mechanical anharmonicity due to through-bond or through-space electro-static interactions; electrical anharmonicity due to nonlinearity of the dipole moment with respect to the vibrational coordinate; and quantum number dependent dephasing dynamics (7). The systematic experimental investigation of the relative importance of these interaction mechanisms is currently lacking, however. Our work is focused on a model system of two coupled vibrations: symmetric and asymmetric CO stretches of rhodium dicarbonyl acetylacetonate (RDC) in hexane (8, 9). Since the linewidth of vibrational transitions (~ 3 cm^{-1}) is small relatively to the strength of inter-mode coupling, all transitions can be easily spectrally resolved providing the intuitive picture of interference between different evolution pathways contributing to the signal. The vibrational energy level structure of the compound is well understood (Figure 1). The interaction between the symmetric (2084 cm^{-1}) and asymmetric (2015 cm^{-1}) C-O stretching motions leads to the anharmonicities of the nuclear potential. The anharmonic frequency shifts, determined from the overtone spectrum, are measured at 11 cm^{-1} and 14 cm^{-1} for the symmetric and asymmetric vibrations, while the frequency of the combination band is red shifted by 24 cm^{-1} with respect to the sum of the fundamental frequencies.

Experimental

Pump-probe experiments are traditionally used to study the population dynamics in condensed phase (10). The population change in the sample is induced by an intense pump pulse. The dynamics, following the excitation pulse, are examined by a time delayed probing pulse. The third-order polarization, induced in the sample by the incident fields, radiates a signal field collinear with

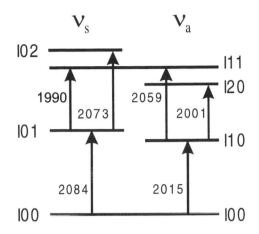

Figure 1. Energies (in cm⁻¹) of singly excited (one-quantum) and doubly excited (two-quantum) vibrational levels of the symmetric (v_s) and asymmetric (v_a) CO vibrations of RDC. All transition frequencies are within the bandwidth of the IR pulses used in the experiment.

the probe beam. If the external fields are resonant with the molecular transitions, the nonlinear response is predominantly dichroic, and the probing field serves as an efficient local oscillator for heterodyne-detection of the signal field (*11*). In addition to population dynamics, pump-probe experiments with ultrashort laser pulses can examine the dynamics of the coherent superpositions of vibrational motions (*12*). The formation of such superposition states is an indication of coupling between the corresponding vibrational motions. Further insight into the mechanisms of vibrational coupling can be gained by frequency dispersing the pump-probe signals. Such experiments reveal the vibrational frequencies sampled by the molecular system and can be used to characterize the potential surface governing the vibrational motion.

The experiment uses 90 fsec near-transform-limited pulses (λ=5μm) obtained by compressing the mid-IR radiation generated by difference frequency mixing of signal and idler outputs of a near-IR BBO OPA in a 1 mm Type II AgGaS₂ crystal. The detailed account of the generation, propagation and compression of mid-IR pulses is given elsewhere (*13*). The pump and probe pulses, obtained by splitting the collimated IR beam, were focussed and crossed in the sample by an f=10 cm parabolic reflector. A stepper motor delay stage was used to control the time between the pump and probe pulses. The probe beam was dispersed in an f=190 mm grating monochromator. The intensity variations of the probe were detected with a liquid nitrogen cooled single channel HgCdTe detector. The polarizations of each of the incident beams, as well as the signal field polarization, were separately controlled with wire grid polarizers.

Results and Discussion

The results of frequency resolved pump-probe experiments are shown on Figure 2. The data was taken by fixing the time delay between the pump and probe pulses to 500 fsec and scanning the frequency of the spectrometer.

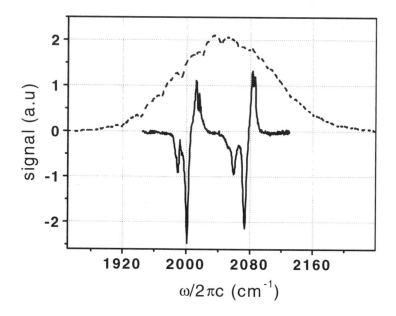

Figure 2. Frequency dispersed pump-probe signal from RDC in hexane (solid line). The delay between the pump and the probe pulses was set to 500 fsec. Dotted trace shows the spectrum of the pulses used in the experiment.

The dispersed data clearly show that the signal is radiated at six different vibrational frequencies. Since the nonlinear response is heterodyne-detected the relative phase of different contributions is preserved. The positive signal, corresponding to bleaching of molecular absorption is observed at the frequencies of fundamental transitions in RDC. The dip in the bleaching signal is due to the absorption of the LO field at the fundamental frequencies. In addition to the bleaching signal, experimental data show four negative peaks, observed at ω =1990, 2001, 2059, and 2073 cm^{-1}.

Third-order nonlinear signal can be described by the system response function, expressed in terms of three-time correlation functions in the dipole moment operator:

$$R^{(3)}_{ijkl} = \left(\frac{i}{\eta}\right)^3 \left\langle \left[\left[\left[\mu_i(t_3), \mu_j(t_2)\right], \mu_k(t_1)\right], \mu_l(0)\right]\right\rangle \tag{1}$$

where $t_i - t_{i-1} = \tau_i$ is a time delay between the incident pulses. Expanding this commutator results in a set of Liouville pathways, which describe the possible evolution paths of the system. The nonlinear response is determined by the interference of different contributing evolution pathways. The example of such interfering pathways is given on Figure 3. Vibrational states of the system are represented by state-vectors $|ab\rangle$, where a and b are the number of quanta in the asymmetric and symmetric stretch, respectively. In Pathways A and C the pump pulse creates a population in $|10\rangle$ state.

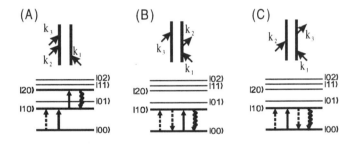

Figure 3. Feynman diagrams and ladder diagrams representing an example of competing pathways contributing to the third-order signals from systems of coupled anharmonic vibrations.

Following the Pathway A, the probe pulse excites a coherence between a singly excited (one-quantum) state and the doubly excited (two-quantum) state. In Pathway B, as well as Pathway C, the final coherence is between the ground state and a one-quantum state. Pathways B and C contribute to the resulting polarization with the same sign and are often referred to as bleaching and stimulated emission signals, respectively. Pathway A, however, is contributing to the total signal with the opposite sign and is referred to as the excited state absorption. It is clear, that in a harmonic system the frequencies of the signal emitted by all three pathways are identical. Furthermore, due to harmonic scaling of the dipole moment $\mu(1 \rightarrow 2) = \sqrt{2}\mu(0 \rightarrow 1)$, the combined contribution of Pathways B and C is exactly canceled by the Pathway A, leading to quenching of nonlinear response. The discussed example of interfering pathways is quite general: third-order nonlinear signal is the result of destructive interference between pathways involving **only** one-quantum states of the system and pathways involving two-quantum states (*14*). If the system is anharmonic the signal due to Pathway A is radiated at a frequency different from the fundamental frequency and the combined contribution of all pathways does not vanish. Figure 2 clearly demonstrates the phase relationship between different contributions to the pump-probe signals. Furthermore, the resonances in the frequency dispersed pump-probe signal can be used to reconstruct the level structure of the one-quantum and two-quantum manifolds of the symmetric and asymmetric CO vibrations in RDC. Four frequencies, contributing to the excited state absorption, correspond to the resonances between the one-quantum and two-quantum states: $|10\rangle \rightarrow |20\rangle$, $|01\rangle \rightarrow |02\rangle$, $|10\rangle \rightarrow |11\rangle$, and $|01\rangle \rightarrow |11\rangle$. In particular, since transitions $|10\rangle \rightarrow |11\rangle$ and $|01\rangle \rightarrow |11\rangle$ share a common upper state, the following assignments can be made: $E(11) - E(10) = 2059$ cm^{-1}, $E(11)-E(01)$ $= 1990$ cm^{-1}, $E(20) - E(1,0) = 2001$ cm^{-1}, and $E(02) - E(01) = 2073$ cm^{-1}. As expected, these frequencies are identical with the previously reported data (Figure 1).

Figure 4 presents the polarization-sensitive frequency-resolved pump-probe data of RDC in chloroform, collected by tuning the spectrometer to the fundamental frequency of the asymmetric CO vibration and scanning the time delay between the pump and probe pulses. The polarization dependence of the pump-probe signal was measured by rotating the polarization of the pump beam with respect to the probe and signal polarizations. The experimentally measured all parallel, R_{ZZZZ}, and perpendicular, R_{ZZYY}, responses were used to calculate the isotropic R_{iso} and anisotropic R_{aniso} responses

$$R_{aniso} = R_{ZZZZ} - R_{ZZYY}$$
$$R_{iso} = (R_{ZZZZ} + 2R_{ZZYY})/3 \tag{2}$$

The signal in all parallel and perpendicular polarization geometries, as well as the anisotropic response, clearly shows the quantum beats at the frequency of 70 cm^{-1}, which corresponds to the formation of the coherent superposition of the symmetric and asymmetric CO vibrations in RDC. Similar to the previous discussion, the observation of the quantum beats in the pump-probe experiment is an indication of the imperfect cancellation of the vibronic pathways involving **only** one-quantum states and pathways involving two-quantum vibrational states. Such imperfect cancellation is due to the red shift of the energy of the combination band, $|11\rangle$ state, relative to the sum of the energies of the

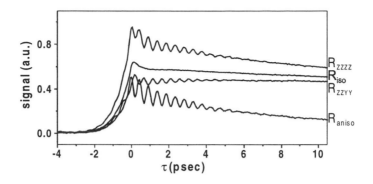

Figure 4. Polarization-sensitive frequency-resolved pump-probe data of RDC in chloroform. The signal is detected at the fundamental frequency of the asymmetric vibration $\omega = 2014$ *cm*$^{-1}$*. Isotropic* R_{iso} *and anisotropic* R_{aniso} *responses were calculated from the experimental data for all-parallel (*R_{ZZZZ}*) and perpendicular (*R_{ZZYY}*) polarization geometries using Eq. 2.*

corresponding fundamental transitions, which results from the interaction of the symmetric and asymmetric CO vibrations.

The formation of coherent superposition of two vibrational motions requires the interactions of the external laser fields with two transition dipole moments, which, in general, can be arbitrarily oriented with respect to each other. This is in contrast with the "population" pathways (see Figure 2), where all interactions occur with the same vibrational motion and the transition dipole moments for the pump, probe and signal emission are, therefore, collinear. Clearly, the mutual orientation of the transition dipole moments of the interacting vibrations will affect the polarization dependence of the nonlinear response. If the contributions of the reorientational and vibronic motions can be separated, based on the difference of the timescales characterizing the two types of motion, then the resulting total response function, describing the overall nonlinear signal, is a sum of products of the orientational and vibronic components:

$$R^{(3)}_{ijkl}(\tau_1,\tau_2,\tau_3) = \sum_{paths} R^{or}_{ijkl}(\tau_1,\tau_2,\tau_3)R^{vib}(\tau_1,\tau_2,\tau_3) \tag{3}$$

where the orientational part of the response function depends on the vibronic pathway. The orientational contributions to the response function can be derived by projecting the polarizations of the incident fields onto the directions of the molecular transition dipoles, averaging over all molecular orientations, and accounting for orientational diffusion between interactions with the external fields (*15*). The transition dipoles of the interacting vibrations can, in general, form an arbitrary angle Θ and are assumed to be fixed in the molecular frame of reference. A full derivation of the orientational response functions for third-order nonlinear spectroscopy will be presented elsewhere (16). Equations 4 present the orientational parts of the third order response function for a system of two vibrations **a** and **b** with mutually orthogonal transition dipole moments, $\Theta = 90°$. Since the orientational component depends on the order of electric field interaction with the dipole moments **a** and **b**, four distinct possibilities can be identified: **aaaa**, **abba**, **abab**, and **aabb**:

$$R_{zzzz}^{aaaa} = \frac{1}{9}c_1(\tau_1)c_1(\tau_3)(1 + \frac{4}{5}c_2(\tau_2))$$

$$R_{yyzz}^{aaaa} = \frac{1}{9}c_1(\tau_1)c_1(\tau_3)(1 - \frac{2}{5}c_2(\tau_2))$$

$$R_{zzzz}^{aabb} = \frac{1}{9}c_1(\tau_1)c_1(\tau_3)(1 - \frac{2}{5}c_2(\tau_2))$$

$$R_{yyzz}^{aabb} = \frac{1}{9}c_1(\tau_1)c_1(\tau_3)(1 + \frac{1}{5}c_2(\tau_2))$$

$$(4)$$

$$R_{zzzz}^{abab} = R_{zzzz}^{abba} = \frac{1}{15}c_1(\tau_1)c_1(\tau_3)c_2(\tau_2)$$

$$R_{yyzz}^{abab} = R_{yyzz}^{abba} = -\frac{1}{30}c_1(\tau_1)c_1(\tau_3)c_2(\tau_2)$$

where τ_1, τ_2, and τ_3 are the respective time delays between the first and second pulses, second and third pulses, and third pulse and the detection time. Coefficients $c(\tau)$ describe the orientational diffusion of the molecule: $c_l(\tau) = \exp(-l(l+1)D_{or}\tau)$, where D_{or} is the orientational diffusion coefficient.

Since quantum beats result from the vibronic pathways where the first two interactions occur with two different vibrations, their polarization dependence is given by R^{abab} and R^{abba} responses. Furthermore, the orientational response of quantum beats between two vibrations with orthogonal transition dipole moments is predicted to be purely anisotropic:

$$R_{iso(beats)}^{abab} = R_{iso(beats)}^{abba} = (R_{zzzz(beats)}^{abba} + 2R_{yyzz(beats)}^{abba})/3 = 0 \qquad (5)$$

Consequently, since the dipole moments of the symmetric and asymmetric CO stretches in RDC are mutually perpendicular, $\Theta = 90°$, the quantum beats are predicted to vanish in the isotropic pump-probe measurement. This is in excellent agreement with the experimental data (Figure 4).

The tensor components of the orientational response function for a system of arbitrarily oriented dipole moments can be calculated by forming the appropriate linear combination of the responses for collinear and orthogonal components of the dipole moments. It is clear from the results in Eq. 4 and the experimental results in Figure 4 that the ratio of amplitudes of the quantum beats for two independent polarization geometries can be used to determine the relative orientation of the interacting dipole moments. For example, if the angle between the dipole moments of two interacting vibrations is Θ, the ratio of the amplitudes of the quantum beats in the anisotropic and isotropic measurements is:

$$\frac{A_{aniso(\,beats\,)}}{A_{iso(\,beats\,)}} = \left(\frac{2}{5} + \frac{3}{10} tan^2 \Theta \right) \tag{6}$$

Frequency dispersed pump probe can provide information about the lifetimes of the vibrational states and can observe the coupling of vibrational motions through the formation of coherent superpositions of the corresponding vibrational states. Generally, however, the signal detected at a certain wavelength can result from several vibrational pathways. In particular, the signal observed at the fundamental frequency of the asymmetric vibration can be due to pathways involving the pump interaction with the symmetric vibration or asymmetric vibration. Direct information about the correlations of different vibrational motion can be gained by 2D IR photon echo spectroscopy (*1-5, 17, 18*). 2D spectroscopy probes the correlations of molecular motions by manipulating vibrational coherences with external light fields. The observable is a 2D map of frequencies sampled by the system during two separate time periods: τ_1 and τ_3. The diagonal features contain information about the nuclear potential governing molecular motion along a single vibrational coordinate. The interaction of different vibrational motions is observed as the formation of anti-diagonal features (cross-peaks) in the 2D spectrum. The shape and the intensity of the cross-peaks relate to the coupling strength and the relative orientation of the interacting structural groups. The nonlinear signal due to three incident laser fields $E_1(k_1)$, $E_2(k_2)$ and $E_3(k_3)$ is generated in $k_{sig}=-k_1+k_2+k_3$ wave-vector matching direction. After the first pulse E_1 the system propagates in a vibrational coherence during time τ_1. Pulses E_2 and E_3, incident simultaneously, can lead to (1) the rephasing of the original coherence, (2) excitation of a coherence within a different vibrational manifold or (3) excitation of higher lying vibrational coherences. The evolution of the system during the final coherence time τ_3 was monitored by spectral interferometry of the signal field with a separate local oscillator (LO) field. The 2D IR data sets were taken by scanning the τ_1 delay for each detection frequency ω_3 and subsequently stepping the detection frequency. 2D IR spectra in ω_1 and ω_3 were obtained by Fourier transforming the experimental data along τ_1 dimension.

Figure 5 presents the absolute value of the polarized (R_{ZZZZ}) heterodyne-detected 2-D photon echo spectrum of RCD in hexane. The diagonal

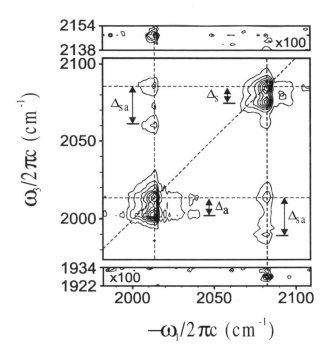

Figure 5. Absolute value R$_{ZZZZ}$ 2D IR-PE spectrum of RDC in hexane. The signal in the upper and lower panels is magnified by a factor of 100 to emphasize the weak spectral peaks at (ω_1, ω_3)=(2015 cm^{-1}, 2148 cm^{-1}) and (2084 cm^{-1}, 1928 cm^{-1}). Dashed lines indicate the frequencies of the fundamental transitions.

contributions to the 2-D spectra characterize the coherent processes occurring within one vibrational manifold. Each diagonal feature consists of two spectral peaks of similar intensities separated along the ω_3 axis by 11 cm^{-1} for symmetric and 14 cm^{-1} for asymmetric vibrations. The relative strength of the two diagonal features is found to depend strongly on the central frequency of the excitation pulse, which was set to 2050 cm^{-1} for this measurement. The anti-diagonal contributions to the 2-D spectrum or cross-peaks represent the processes involving two different vibrational transitions and therefore contain information pertaining to the coupling of the corresponding vibrational motions. Similarly to the diagonal features each of the anti-diagonal features is composed of two spectral peaks, which are separated along the ω_3 axis by 24 cm^{-1}. The amplitude of the anti-diagonal features is ~1/3 of the amplitude of the diagonal contributions. In addition to the eight relatively strong diagonal and anti-diagonal peaks, two single weak peaks were observed at $(\omega_1, \omega_3)=(2015 \text{ cm}^{-1}, 2148 \text{ cm}^{-1})$ and $(2084 \text{ cm}^{-1}, 1928 \text{ cm}^{-1})$ (See inserts on Figure 5).

The observed structure of the 2D spectrum is easily understood by considering the interference of evolution pathways involving one-quantum and two-quantum states. For example, the diagonal spectral feature at $(\omega_1, \omega_3)=(2015 \text{ cm}^{-1}, 2000\text{-}2015 \text{ cm}^{-1})$ is due to the evolution pathways described in Fig. 3. The spectral peak, appearing exactly on the diagonal of the 2D spectrum $(\omega_1, \omega_3)=(2015 \text{ cm}^{-1}, 2015 \text{ cm}^{-1})$, represents the processes during which the system evolves at the fundamental frequency of the asymmetric stretching mode during both τ_1 and τ_3 time periods (pathways B and C). On the other hand, the spectral peak, appearing immediately below the diagonal $(\omega_1, \omega_3)=(2015\text{cm}^{-1}, 2001\text{cm}^{-1})$ is due to the pathway involving a two-quantum state (Pathway A, Figure 3). Following Pathway A, during the first time period, τ_1, the system evolves at the fundamental frequency of the asymmetric vibration, while during the final time, τ_3, the evolution is at the difference frequency between the one-quantum and two-quantum states within the asymmetric stretch vibrational manifold. The separation of two spectral peaks constituting a diagonal feature is therefore directly related to the anharmonicity of the nuclear potential.

It is clear from the above discussion that 2D spectrum lends itself to immediate and highly intuitive interpretation and assignment. A single 2D spectrum of RDC carries complete information about the level structure of the singly and doubly excited states of the symmetric and asymmetric stretch manifolds. The anharmonic frequency shifts, $\Delta_s=11 \text{ cm}^{-1}$ for the symmetric and $\Delta_a=14 \text{ cm}^{-1}$ for the asymmetric stretch, are measured directly from the separation of the spectral peaks constituting the corresponding diagonal features. The $\Delta_{sa}=24 \text{ cm}^{-1}$ splitting of the peaks in the anti-diagonal features represents the red shift of the combination band frequency with respect to the sum of fundamental frequencies and is related to the strength of the coupling between the symmetric and asymmetric C-O vibrational modes. One of the finer demonstrations of the sensitivity of the 2D spectroscopy to the anharmonicities of the nuclear potential is the observation of the weak spectral features at $(\omega_1, \omega_3)=(2015 \text{ cm}^{-1}, 2148\text{cm}^{-1})$ and $(2084 \text{ cm}^{-1}, 1928 \text{ cm}^{-1})$. These peaks are due to processes involving three-quantum transitions between different vibrational manifolds, which are strictly forbidden in a harmonic system.

The anti-diagonal features of 2-D spectra represent the pathways where the polarization is evolving at two different vibrational frequencies during the τ_1 and τ_3 time periods. Following the previous discussion, the dependence of the diagonal and antidiagonal features on the polarization of the incident fields is expected to be different. The simulation of the 2-D spectrum (not shown) demonstrates that the 3/1 ratio of the experimental intensities of the diagonal and anti-diagonal features is due to the difference of the orientational parts of the response functions. Similarly to the pump-probe experiments, the measurement of the 2D spectra in two different polarization geometries can be used to determine the angle between the interacting dipoles. An example of such experiment is the measurement of the anisotropy of the nonlinear response (6). The anisotropy value, defined as the ratio of the anisotropic to the isotropic responces:

$$I(\omega_1,\omega_3) = (R_{ZZZZ} - R_{ZZZZ})/(R_{ZZZZ} + 2R_{ZZZZ}) \tag{7}$$

is expected to equal 2/5 for the diagonal features in the 2D spectrum. The anisotropy of the off-diagonal features, however, will depend on the angle between the transition dipole moments of the interacting vibrations and can be used to determine the mutual orientation of the interacting structural groups. Furthermore, as a rephasing experiment, 2D photon-echo spectroscopy can characterize the inhomogeneous ensemble of interacting vibrations. A 2D map of anisotropy will, therefore, directly measure the correlation between the distribution of vibrational frequencies and the distribution of mutual orientations of the constituting structural groups.

Eq. 4 provide a general framework for calculating the orientational part of the third-order response function in a system of coupled vibrations. These expressions allow the determination of the projection angle between the corresponding transition dipole moments from the measurement of two independent tensor elements of third-order response. The underlying assumption for such treatment is the ability to separate the reorientational and vibronic components based on the difference of timescales for these motions. While generally a good approximation, such approach does not account for possible effects of population transfer and coherence transfer between the interacting motions. For example, the timescale for population transfer between the symmetric and asymmetric CO stretches of RDC in hexane was determined to be 6 psec (8), which is similar to the rates of reorientational diffusion. Although the polarization sensitive pump-probe data of CO vibrations in RDC do not show any discrepancies with the predicted behavior, understanding the effects of population and coherence transfer can prove important for developing 2D third-order spectroscopy into an effective structural tool.

This work was supported by the Office of Basic Energy Sciences, U.S. Department of Energy (DE-FG02-99ER14988), the donors to the ACS-Petroleum Research Fund, and an award by the Research Corporation.

References

1. P. Hamm, M. Lim, W. F. DeGrado and R. M. Hochstrasser, *Proc. Nat. Acad. Sci. USA* **1999,** 96, 2036.
2. S. Mukamel, *Ann. Rev. Phys. Chem.* **2000,** 51, 691.
3. A. Tokmakoff, M. J. Lang, D. S. Larsen, G. R. Fleming, V. Chernyak and S. Mukamel, *Phys. Rev. Lett.* **1997,** 79, 2702.
4. M. C. Asplund, M. T. Zanni and R. M. Hochstrasser, *Proc. Natl. Acad. Sci. USA* **2000,** 97, 8219.
5. W. Zhao and J. C. Wright, *J. Am. Chem. Soc.* **2000,** 84, 1411.
6. S. Woutersen and P. Hamm, *J. Phys. Chem. B* **2000,** 104, 11316.
7. M. Khalil and A. Tokmakoff, *Chem. Phys.* **2001,** 266, 213.
8. J. D. Beckerle, M. P. Casassa, R. R. Cavanagh, E. J. Heilweil and J. C. Stephenson, *Chem. Phys.* **1992,** 160, 487.
9. K. D. Rector, A. S. Kwok, C. Ferrante, A. Tokmakoff, C. W. Rella and M. D. Fayer, *J. Chem. Phys* **1997,** 106.
10. S. Mukamel, *Principles of Nonlinear Optical Spectroscopy* (Oxford University Press, New York, 1995).
11. L. D. Ziegler, R. Fan, A. E. Desrosiers and N. F. Scherer, *J. Chem. Phys.* **1994,** 100, 1823.
12. M. Mitsunaga and C. L. Tang, *Phys. Rev. A* **1987,** 35, 1720.
13. N. Demirdöven, O. Golonzka, M. Khalil and A. Tokmakoff, *Opt. Lett.* **2000,** in preparation.
14. W. M. Zhang, V. Chernyak and S. Mukamel, *J. Chem. Phys.* **1999,** 110, 5011.
15. A. Tokmakoff, *J. Chem. Phys.* **1996,** 105, 1.
16. O. Golonzka and A. Tokmakoff, *J. Chem. Phys.* **2001,** in press.
17. J. D. Hybl, A. W. Albrecht, S. M. Gallager-Faeder and D. M. Jonas, *Chem. Phys. Lett.* **1998,** 297, 307.
18. O. Golonzka, M. Khalil, N. Demirdoven and A. Tokmakoff, *Phys. Rev. Lett.* **2000,** 86, 2154.

Water

Chapter 14

Resonant Intermolecular Energy Transfer in Liquid Water

S. Woutersen and H. J. Bakker

FOM-Institute for Atomic and Molecular Physics, Kruislaan 407, 1098 SJ Amsterdam, The Netherlands

We report on the observation of resonant vibrational energy transfer in a liquid at room temperature. This resonant energy transfer process is observed to occur for the O-H stretch vibration in solutions of HDO in D_2O and in neat liquid water (H_2O). For solutions of HDO in D_2O, the energy-transfer can be quantitatively described with a model in which the O-H groups are coupled via dipole-dipole interactions. Surprisingly, this description strongly underestimates the rate of energy transfer that is observed for neat liquid water, which indicates the presence of additional interaction mechanisms.

Introduction

In recent years it has become possible to measure the dynamics of water molecules directly in the time domain with ultrashort mid-infrared laser pulses. The use of these pulses in nonlinear optical experiments has led to new insights in the structure and dynamics of neat liquid water. For instance, it was found that the reorientational motion of water shows a bimodal distribution that is strongly correlated to the strength of the hydrogen-bond interactions (*1*). In other experiments employing subpicosecond mid-infrared pulses, information was obtained on the hydrogen-bond stretching dynamics of liquid water (*2–4*) and on the energy relaxation of the O-H stretch vibration of the water molecules (*5, 6*). These previous experiments all probed the dynamics of the O-H stretch vibrational mode of HDO molecules dissolved in D_2O instead of the dynamics of the O-H stretch modes of 'real' water (liquid H_2O). Only very recently, two studies of the O-H stretch vibrations of H_2O molecules were reported (*7, 8*). Clearly, a dilute solution of HDO in D_2O, for which the OH oscillators are more or less isolated, can only be regarded as a good model system for liquid water if there is no resonant intermolecular energy transfer between the OH oscillators in liquid H_2O.

Resonant intermolecular energy transfer via Förster energy transfer (9) is a well-known, often observed process for electronic excitations (10–13). For vibrations in the condensed phase, resonant intermolecular energy transfer is a less likely process, because the timescale of vibrational relaxation (picoseconds) is in general much shorter than that of resonant intermolecular energy transfer (nanoseconds) (14). Hence, up to now resonant vibrational energy transfer has only been observed for systems with relatively long vibrational lifetimes like diatomics under cryogenic conditions (14, 15) and specific solid-state systems (16, 17). Interestingly, the time scale of non-resonant vibrational energy transfer can be much shorter than that of resonant vibrational energy transfer (18, 19). Non-resonant vibrational energy transfer has been observed for polyatomic molecular liquids at room temperature (18, 19).

It was found previously for a dilute solution of HDO in D_2O that the OH-stretch excitation rapidly relaxes to the hydrogen bond with a time constant of 740 ± 25 femtoseconds (6). Hence, for resonant intermolecular energy transfer to compete with vibrational relaxation, the interaction between the OH-modes on different water molecules should be very strong,

Experiment

We investigate the occurence of resonant intermolecular energy transfer of an OH-stretch vibrational excitation in solutions of HDO in D_2O and in neat liquid water. In general, the OH-groups of two molecules in liquid water will not be oriented parallel, and transfer of the OH-stretch excitation from one molecule to another will result in a change of the polarization of the excitation. This means that the energy transfer leads to a decay of the rotational anisotropy of the excitation. Hence, intermolecular transfer of the OH-stretch excitation can be measured by exciting a fraction of the OH-oscillators and probing the decrease of the anisotropy of this excitation with increasing time. Since the transfer rate depends strongly on the distance between the 'donor' and 'acceptor' molecule, the average transfer rate will depend strongly on the concentration of OH-oscillators. Hence, by varying the concentration, the direct energy transfer can be distinguished from other contributions to the decay of the rotational anisotropy, in particular the orientational diffusion of the OH-groups.

In the experiment we use two independently tunable mid-infrared pulses with a duration of approximately 200 fs. These pulses are generated via parametric generation and amplification in BBO and KTP crystals. These parametric processes are pumped with the pulses delivered by a regenerative Ti:sapphire laser that have an energy of 1 mJ per pulse, a pulse duration of 120 femtoseconds and a central wavelength of 800 nm. The repetition rate of the system is 1 kHz. The generated mid-infrared pulses are tunable between 2.7 and 3.4 μm.

One of the generated mid-infrared pulses pulses has an energy of approximately 20 μJ and is used as pump pulse, the other has an energy of less than 1 μJ and is used as probe. The pump pulse is tuned to the center frequency of the $v_{OH} = 0 \to 1$ absorption band, and induces a significant population of the $v_{OH} = 1$ level. This excitation results in transient anisotropic absorption changes, which are monitored by the probe pulse that is tuned to the $v_{OH} = 1 \to 2$ transition frequency. The probe polarization is at $45°$ with respect to the pump polarization, and the relative transmission changes $\ln(T/T_0)_{\parallel}$ and $\ln(T/T_0)_{\perp}$ of the parallel and perpendicular polarization components are monitored simultaneously as a function of delay. In this way, both the rotation-free signal $\ln(T/T_0)_{\parallel} + 2\ln(T/T_0)_{\perp}$ and the rotational anisotropy R are determined as a function of the delay between pump and probe. This rotational anisotropy is given by:

$$R = \frac{\ln(T/T_0)_{\parallel} - \ln(T/T_0)_{\perp}}{\ln(T/T_0)_{\parallel} + 2\ln(T/T_0)_{\perp}} \tag{1}$$

Due to the division by the rotation-free signal, R is only sensitive to the orientational dynamics and not to the population relaxation (20).

The samples consist of a thin layer of either HDO dissolved in D_2O or pure H_2O. The HDO:D_2O samples are prepared by mixing appropriate amounts of H_2O (HPLC grade) and D_2O (>99.9 atom% D), and are kept between two CaF$_2$ windows separated by a teflon spacer of 15 or 25 μm thickness. Samples consisting of pure H_2O are prepared by squashing a droplet of H_2O (HPLC grade) between two CaF$_2$ windows, and applying pressure until the sample is sufficiently thin. The sample thickness achieved in this way is approximately 1 μm, as determined from the sample transmission and literature values (21, 22) of the infrared extinction coefficient of liquid H_2O. All experiments are performed at room temperature.

In order to perform pump-probe spectroscopy in transmission, the sample has to be sufficiently thin to transmit a measurable fraction of the probe intensity. Using highly concentrated HDO:D_2O and pure H_2O samples, the required thickness varies from 1 to 25 μm. At these high densities of OH-groups, the vibrational relaxation process (conversion of the vibrational energy into heat) results in a significant temperature increase in the focus. In fact, even a single laser shot will result in a temperature increase of several K. This means that it is essential to avoid accumulated heating in the focus. In previous studies on dilute HDO:D_2O solutions this has been done by circulating the water. However, if the sample is very thin, very high pressures would be needed to circulate the sample. Therefore, to avoid accumulated heating effects, the entire sample is rapidly rotated using a small DC motor.

Results and discussion

General considerations

Fig. 1 represents a typical pump-probe scan, recorded in an 8.7 M solution of HDO in D_2O, with a pump frequency of 3400 cm^{-1} and a probe frequency of 3000 cm^{-1}. The relative transmission changes $\ln(T/T_0)_\parallel$ and $\ln(T/T_0)_\perp$ of the

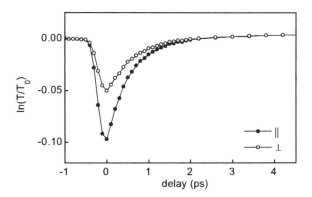

FIGURE 1. Pump-probe scans recorded in liquid HDO:D_2O with $\nu_{pu} = 3400$ cm^{-1} and $\nu_{pr} = 3150$ cm^{-1}, showing the absorption increase at the probe frequency vs. the delay between the pump and probe pulses, for the parallel (solid circles) and perpendicular (open circles) components of the probe pulse.

probe polarization components parallel and perpendicular to the pump polarization are plotted as a function of delay between pump and probe. The negative value of $\ln(T/T_0)$ observed for small delay values is due to the excited-state ($\nu_{OH} = 1 \rightarrow 2$) absorption, and decays as the excited HDO molecules relax to the vibrational ground state. For large delay, $\ln(T/T_0)$ approaches a small positive value due to the temperature increase that occurs upon vibrational relaxation. Both in HDO:D_2O (23) and in pure H_2O (21, 22), the $\nu_{OH} = 0 \rightarrow 1$ absorption band shifts towards higher frequency and decreases in magnitude with increasing temperature. Since the spectrum of the probe pulse slightly overlaps with the red side of the $\nu_{OH} = 0 \rightarrow 1$ absorption band, the increase in temperature upon vibrational relaxation causes a small increase of the transmission of the probe pulse. In order to study the orientational and vibrational relaxation in more detail, we subtract the small transmission increase caused by the temperature rise.

HDO dissolved in D_2O

The rotation-free signal observed for three different HDO concentrations is shown in Fig. 2. For delay values larger than 1 ps, all transients can be described with a single-exponential decay with a time constant of 720 fs. This value represents the vibrational lifetime T_1 of the OH-stretch mode (6). Clearly, the vibrational relaxation dynamics of the OH-stretch mode of HDO in D_2O shows no significant dependence on the HDO concentration.

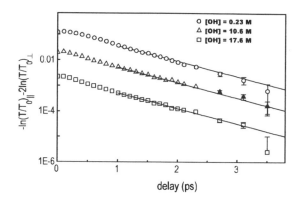

FIGURE 2. Rotation-free excited-state absorption $-\ln(T/T_0)_\parallel$ $-2\ln(T/T_0)_\perp$ recorded in HDO:D_2O solutions with three different HDO concentrations. The pump frequency and probe frequencies are $v_{pu} = 3400$ cm^{-1} and $v_{pr} = 3150$ cm^{-1}, respectively. The transients for [HDO] $= 8.7$ M and [HDO] $= 13$ M have been vertically displaced for clarity. The solid lines represent a single-exponential decays with a time constant of 720 fs.

From the transmission changes $\ln(T/T_0)_\parallel$ and $\ln(T/T_0)_\perp$ of the parallel and perpendicular components of the probe pulse, we can calculate the rotational anisotropy R. Fig. 3 shows R, normalized with respect to the value at zero delay, as a function of the delay between pump and probe. In contrast to the vibrational relaxation, the orientational relaxation dynamics of the OH-stretch excitation clearly depend strongly on the HDO concentration. The contribution of the orientational diffusion of the HDO molecules to the decay of R will *not* depend on the HDO concentration. Hence, the observation of a concentration dependence of the orientational dynamics therefore constitutes strong evidence for the occurence of resonant transfer of the OH-stretch excitation between the HDO molecules (Förster energy transfer (9)). The rate of Förster transfer by dipole-dipole coupling between an excited molecular oscillator and an oscillator in the ground state is proportional to $1/r^6$, where r is the distance between the two oscillators. Since the 'acceptor' transition dipole will generally not be oriented parallel to the 'donor' transition dipole, the Förster transfer leads to a decay of the rotational anisotropy. Because of the strong dependence of the transfer rate on the intermolecular distance, Förster transfer causes the rotational anisotropy to decay faster with increasing concentration, as observed in Fig. 3 (7). To our knowledge, this constitutes the first observation of vibrational Förster energy transfer in a liquid at room temperature.

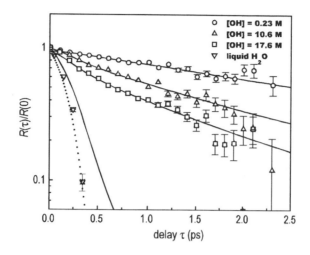

FIGURE 3. Rotational anisotropy R, normalized with respect to the value at delay zero, recorded in HDO:D$_2$O solutions with three different HDO concentrations and in pure H$_2$O. The pump frequency and probe frequencies are $\nu_{pu} = 3400$ cm^{-1} and $\nu_{pr} = 3150$ cm^{-1}, respectively. The solid curves have been calculated using Eq. (8). The dotted curve has been calculated using Eq. (9).

In a quantitative analysis of the data, both the contributions of the random orientational motion of the HDO molecules and the Förster energy transfer of the OH-stretch excitation should be accounted for. In a previous study on the orientational diffusion of water molecules it was found that the random orientational motion of the HDO molecules takes place on two distinct time scales, a slow time scale associated with the reorientation of strongly hydrogen-bonded molecules and a fast time scale associated with weakly hydrogen-bonded molecules (1). The amplitude of the slow component was observed to be much larger than the amplitude of the fast component (24). In these previous experiments, the frequency of the probe pulse was tuned to the $\nu_{OH} = 0 \rightarrow 1$ transition, and consequently the orientational dynamics of both molecules in the vibrational ground ($\nu_{OH} = 0$) state and molecules in the first excited ($\nu_{OH} = 1$) state were probed. In the present experiments, the probe frequency is tuned to the $\nu_{OH} = 1 \rightarrow 2$ transition, and only the excited ($\nu_{OH} = 1$) molecules are probed. These excited molecules are more strongly hydrogen bonded than the molecules in the $\nu_{OH} = 0$ state (25, 26), so that the contribution of the fast component associated with the weakly hydrogen-bonded molecules becomes even smaller than in the experiments in which the $v_{OH} = 0 \rightarrow 1$ was probed (27). As a consequence, we mainly observe the slow time scale of the orientational motion so that the decay of the anisotropy can be described well with a single exponential with decay constant τ_{or}.

In a model often used to describe Förster energy transfer (13, 28), it is assumed that once the excitation has been transferred from an optically excited ($v_{OH} = 1$) molecule to a randomly oriented acceptor ($v_{OH} = 0$) molecule, it no longer contributes to the rotational anisotropy. Detailed calculations show that after one or more transfers the excitation will contribute less than 3% to the rotational anisotropy (28). Hence the contribution of an OH-stretch excitation to the rotational anisotropy will decay as the probability that it is still on the initially excited HDO molecule. If a particular molecule is excited at $t = 0$, then the probability $\rho(\tau)$ that at $t = \tau$ the molecule is still excited is given by (29)

$$\rho(\tau) = \langle \prod_{j=1}^{N} \exp[-w_j\tau] \rangle, \tag{2}$$

where w_j is the rate constant for Förster energy transfer from the excited molecule to acceptor molecule j, N is the number of acceptor molecules, and $\langle \ldots \rangle$ denotes a statistical average over all possible spatial distributions of the acceptor molecules. The assumption that once the excitation has been transferred, it no longer contributes to the rotational anisotropy implies that $R(\tau)$ is proportional $\rho(\tau)$. If we assume for simplicity that w_j depends only on the distance r_j between the donor and acceptor (not on their relative orientation), we have the well-known relation

$$w_j = \frac{1}{T_1}\left(\frac{r_0}{r_j}\right)^6, \tag{3}$$

where r_0 is a constant, usually referred to as the Förster radius (29, 30). The acceptor molecules are randomly distributed, so the probability of finding acceptor j at a distance between r_j and $r_j + dr_j$ is simply $4\pi r_j^2 dr_j/V$, with V the total volume of the sample. Averaging over this probability distribution for each molecule, we have

$$\rho(\tau) = \left\{ \frac{4\pi}{V} \int_0^{\Lambda} \exp\left[-\frac{\tau}{T_1}\left(\frac{r_0}{r}\right)^6\right] r^2 dr \right\}^N, \tag{4}$$

where we have assumed that the sample is a sphere with radius Λ, centered at the excited molecule (hence $V = 4\pi\Lambda^3/3$). Performing the integration, we obtain

$$\rho(\tau) = \left\{ \exp\left[-\frac{\tau}{T_1}\left(\frac{r_0}{\Lambda}\right)^6\right] + \sqrt{\frac{\pi\tau}{T_1}\left(\frac{r_0}{\Lambda}\right)^6} \left[\mathrm{erf}\left(\sqrt{\frac{\pi\tau}{T_1}\left(\frac{r_0}{\Lambda}\right)^6}\right) - 1 \right] \right\}^N. \tag{5}$$

Using the fact that the concentration (expressed in particles per unit of volume) is $[C] = 3N/4\pi\Lambda^3$ to eliminate Λ, and expanding in terms of $1/N$, we obtain

$$\rho(\tau) = \left\{ 1 - \frac{4\pi^{3/2}[C]}{3}\sqrt{\frac{r_0^6\tau}{T_1}}\left(\frac{1}{N}\right) + O\left(\frac{1}{N}\right)^2 \right\}^N . \tag{6}$$

Taking the thermodynamic limit $N \to \infty$, we obtain

$$\rho(\tau) = \exp\left(-\frac{4\pi^{3/2}}{3}[C]\sqrt{\frac{r_0^6\tau}{T_1}} \right). \tag{7}$$

The delay and concentration dependence given by Eq. (??) for the decay of the rotational anisotropy has been experimentally confirmed for optical excitations in the visible (11, 12).

If we assume that the orientational motion of the HDO molecules and the Förster transfer are independent processes, the total decay of the anisotropy is given by:

$$\frac{R(\tau)}{R(0)} = \exp\left(-\frac{\tau}{\tau_{or}} - \frac{4\pi^{3/2}}{3}[HDO]\sqrt{\frac{r_0^6\tau}{T_1}} \right). \tag{8}$$

We found that all scans recorded in HDO:D_2O can be well described by Eq. (8) with $\tau_{or} = 4.0$ ps and a Förster radius $r_0 = 2.1$ Å (solid curves in Fig. 3). From Eq. (3) we see that r_0 is the distance between two OH-oscillators at which the Förster transfer rate becomes comparable to the vibrational relaxation rate.

Liquid H$_2$O

Fig. 3 also shows the decay of the rotational anisotropy observed for the OH-stretch excitation in liquid H_2O. This decay occurs on a very rapid time scale, only slightly slower than the cross-correlation trace $C(\tau)$ of the pump and probe intensity envelopes. To investigate this in more detail, we compare both the numerator $\ln(T/T_0)_{\parallel} - \ln(T/T_0)_{\perp}$ and the denominator $\ln(T/T_0)_{\parallel} + 2\ln(T/T_0)_{\perp}$ of R with the cross-correlation function $C(\tau)$, as shown in Fig. 4. Clearly, the denominator decays more slowly than the cross-correlation function, which implies that the population relaxation, and hence the decay of $\ln(T/T_0)_{\parallel}$ and $\ln(T/T_0)_{\perp}$, does not take place very fast. In contrast, the numerator $\ln(T/T_0)_{\parallel} - \ln(T/T_0)_{\perp}$ does not differ significantly from the cross-correlation. Since the decay of $\ln(T/T_0)_{\parallel}$ and $\ln(T/T_0)_{\perp}$ is much slower (on the order of that of the rotation-free signal), this must be due to a fast orientational scrambling of the OH-stretch excitation that rapidly reduces the difference between $\ln(T/T_0)_\parallel$ and $\ln(T/T_0)_\perp$ to zero. Note however, that the anisotropy decay is not instantaneous because in that case the initial value of R would have been zero.

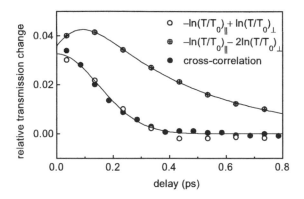

FIGURE 4. Numerator $\ln(T/T_0)_\parallel - \ln(T/T_0)_\perp$ and denominator $\ln(T/T_0)_\parallel + 2\ln(T/T_0)_\perp$ of the rotational anisotropy. The denominator has been scaled by a factor 1/6 for better comparison. Also shown is the cross-correlation of the intensity envelopes of the pump and probe pulses. The solid lines show least-squares fits to the cross-correlation (Gaussian) and to the denominator (convolution of a Gaussian with a single-exponential decay).

Curiously, even though the difference $\ln(T/T_0)_\parallel - \ln(T/T_0)_\perp$ follows the cross-correlation $C(\tau)$, the decay of $R(\tau)$ can nevertheless still take place *slower* than that of $C(\tau)$. This can be understood as follows. If the orientational relaxation is significantly faster than the duration of the pulses, the delay dependence of the numerator of R is given by $C(\tau)$, and we have

$$R(\tau) \propto \frac{C(\tau)}{\ln[T(\tau)/T_0]_\parallel + 2\ln[T(\tau)/T_0]_\perp}. \tag{9}$$

If the denominator decays on a fast enough time scale, it will decrease significantly with increasing τ even within the duration of the cross-correlation. In that case, $R(\tau)$ is given by $C(\tau)$ divided by a rapidly decaying function, and hence will decay slower than $C(\tau)$. Only if the denominator is a constant (infinitely slow decay of the rotation-free signal) R will decay equally fast as $C(\tau)$. Thus, the decay of R in H_2O observed in Fig. 3 does not mirror the orientational relaxation, but is determined by the cross-correlation function and the decay of the rotation-free signal (the denominator of R), which is determined by population relaxation only.

Indeed, if we evaluate (9) using least-squares fits to the cross-correlation function and rotation-free signal (solid curves in Fig. 4), we find that this expression gives a reasonable description of decay of the rotational anisotropy in H_2O (dotted curve in Fig. 3). This shows that in liquid water the orientational scrambling of the OH-stretch excitation occurs on a time scale shorter than the time resolution of our experiments (~ 100 fs).

The solid curve in Fig. (3) that is closest to the dotted curve is calculated with Eq. (8) using the values for τ_{or} ps and r_0 obtained from the least-squares fit to the HDO:D_2O data and an OH concentration of 111 M, which is twice the concentration of H_2O molecules in liquid H_2O at room temperature. Clearly, the rate of energy transfer predicited by the dipole-dipole interaction mechanism is significantly slower than the observed rate.

Part of the orientational scrambling observed in H_2O will be due to intramolecular transfer of the OH-stretch excitation. Excitation of either the symmetric or antisymmetric OH-stretch mode of the H_2O molecule is followed by a rapid intramolecular redistribution of the excitation over the two modes (31). It is easily shown that the complete transfer of a dipole excitation from an initially excited oscillator to an accepting oscillator changes the rotational anisotropy from the initial value of $\frac{2}{5}$ to $\frac{2}{5}P_2(\cos\delta)$, where P_2 is the second-order Legendre polynomial and δ the angle between the two transition dipole moments (32). Since for the symmetric and antisymmetric OH-stretch modes of H_2O we have $\delta = 90°$, and the excitation will be distributed with equal probabilities over the symmetric and antisymmetric modes, we expect that the intramolecular redistribution will reduce the anisotropy to $\frac{1}{2}(\frac{2}{5} + \frac{2}{5}P_2(0)) = \frac{1}{10}$. Thus, the intramolecular transfer causes the rotational anisotropy to decrease to a quarter of its initial value. The complete vanishing of R is caused by the intermolecular energy transfer.

It should be noted that for neat liquid water the intermolecular energy transfer will be of a more complicated nature than is described by Eq. (3). Because of the high density of OH-groups in liquid H_2O, a simple expression of the form of Eq. (3), which describes the interaction between two point dipoles at a large distance, can no longer be used. In fact, the OH-groups in water are so close to each other that not only dipole-dipole, but also dipole-quadrupole, quadrupole-quadrupole, and higher-order interactions will become important (33). The Förster expression hinges on the fact that these other interactions decrease much more rapidly with increasing distance than the dipole-dipole interaction, and thus become less important. At the short distances occurring in H_2O, this is of course not the case, and a more general treatment of the intermolecular coupling is necessary to accurately describe this liquid. Another effect that could lead to a break-down of the description given by Eq. (8) is that for liquid H_2O it will no longer be correct to assume that the acceptor molecules are randomly distributed. The hydrogen bonds in liquid water will lead to a local ordering which will also affect the rate of energy transfer. In addition to the fact that Eq. (8) is no longer appropriate in describing the electrical interactions of the OH groups in neat liquid water, there might also be other types of interactions that contribute to the intermolecular energy transfer. For instance, for water molecules that are bound by a strong hydrogen bond, the OH groups may be coupled by intermolecular anharmonic interactions through the hydrogen bond.

For neat liquid water, the intermolecular energy transfer of the O-H excitation is much faster than the vibrational relaxation, i.e. relaxation to other (lower) energy degrees of freedom. This means that the vibrational energy is transferred by many water molecules before dissipation occurs. The rapid energy transfer among water molecules also means that OH groups in a hydrophobic environment will remain much longer in a vibrationally excited state than OH groups in a hydrophilic environment.

Conclusions

We have studied the decay of the rotational anisotropy of the OH-stretch excited-state absorption in water as a function of the concentration of OH-oscillators. In $HDO:D_2O$, the decay rate increases significantly with increasing HDO concentration. This constitutes evidence for rapid intermolecular transfer of the OH-stretch excitation. We found that the transfer can be described quantitatively by a dipole-dipole energy (Förster) transfer mechanism.

In H_2O, the orientational scrambling of the OH-stretch excitation occurs on a time scale that is shorter than the time resolution of our experiment ($\sim$100 fs). This rapid orientational scrambling is caused by both intramolecular redistribution of the OH-stretch excitation over the symmetric and asymmetric OH-stretch modes of the H_2O molecules, which reduces the anisotropy to a quarter of its initial value, and a very rapid intermolecular transfer of the OH-stretch excitation between the H_2O molecules. The intermolecular energy transfer probably involves higher-order electrical interactions and intermolecular anharmonic interactions, possibly through the hydrogen bonds that connect the H_2O molecules.

Acknowledgment

The research presented in this paper is part of the research program of the Stichting Fundamenteel Onderzoek der Materie (Foundation for Fundamental Research on Matter) and was made possible by financial support from the Nederlandse Organisatie voor Wetenschappelijk Onderzoek (Netherlands Organization for the Advancement of Research).

References

1. Woutersen, S.; Emmerichs, U.; Bakker, H. J., Femtosecond mid-infrared pump-probe spectroscopy of liquid water: evidence for a two-component structure. *Science* **1997** *278*, 658–660.

2. Laenen, R.; Rauscher, C.; Laubereau, A., Dynamics of local substructures in water observed by ultrafast infrared hole burning. Phys. Rev. Lett. **1998**, *80*, 2622–2625.

3. Gale, G. M.; Gallot, G.; Hache, F.; Lascoux, N.; Bratos, S.; Leicknam, J.-C., Femtosecond Dynamics of Hydrogen Bonds in Liquid Water: A Real Time Study. *Phys. Rev. Lett.*, **1999**, *82*, 1068–1071.

4. Woutersen S.; Bakker, H. J., Hydrogen bond in liquid water as a Brownian oscillator. *Phys. Rev. Lett.*, **1999**, *83*, 2077–2080).

5. Graener, H.; Seifert, G.; Laubereau, A., New spectroscopy of water using tunable picosecond pulses in the infrared. *Phys. Rev. Lett.* **1991**, **66**, 2092–2095.

6. Woutersen, S.; Emmerichs, U.; Nienhuys, H.-K.; Bakker, H. J., Anomalous temperature dependence of vibrational lifetimes in water and ice. *Phys. Rev. Lett.* **1998**, *81*, 1106–1109.

7. Woutersen S.; Bakker, H. J., Resonant intermolecular transfer of vibrational energy in liquid water. *Nature*, **1999**, *402*, 507–509.

8. Deak, J.; Rhea, S.; Iwaki, L.; Dlott, D., Vibrational energy relaxation and spectral diffusion in water and deuterated water. *J. Phys. Chem.* **2000**, *A104*, 4866–4874.

9. Förster, T., Transfer mechanisms of electronic excitation. *Discussions Faraday Soc.* **1959**, *27*, 7–17.

10. Bennet, R. G.; Schwenker, R. P.; Kellogg, R. E., Radiationless intermolecular energy transfer. I. Singlet→singlet transfer. *J. Chem. Phys.* **1964**, *41*, 3037–3040.

11. Eisenthal, K. B., Measurement of intermolecular energy transfer using picosecond light pulses. *Chem. Phys. Lett.* **1970**, *6*, 155–157.

12. Rehm D.; Eisenthal K. B., Intermolecular energy transfer studied with picosecond light pulses. *Chem. Phys. Lett.* **1971**, *9*, 387–389.

13. Drake, J. M.; Klafter, J.; Levitz, P., Chemical and biological microstructures as probed by dynamic processes. *Science* **1991**, *251*, 1574–1579.

14. Vlahoyannis, Y.P.; Krueger, H.; Knudtson, J. T.; Weitz, E. Vibration-vibration energy transfer processes in liquid xenon: a measurement of the rate constant for HCl(v=2) + HCl(v=0) → 2HCl(v=1). *Chem. Phys. Lett.* **1985**, *121*, 272-278.

15. Anex, D. S.; Ewing, G. E., Transfer and storage of vibrational energy in liquids: Collisional up-pumping of carbon monoxide in liquid argon. *J. Phys. Chem.* **1986**, *90*, 1604–1610

16. Morin, M.; Jakob, P.; Levinos, N. J.; Chabal, Y. J.; Harris, A. L., Vibrational energy transfer on hydrogen-terminated vicinal Si(111) surfaces: interadsorbate energy flow. *J. Chem. Phys.* **1992**, *96*, 6203–6212.

17. Brugmans, M. P. J.; Bakker, H. J.; Lagendijk, A., Direct vibrational energy transfer in zeolites. *J. Chem. Phys.* **1996**, *104*, 64–84.

18. Ambroseo, J. R.; Hochstrasser, R. M., Pathways of relaxation of the N-H stretching vibration of pyrrole in liquids. *J. Chem. Phys.*, **1988**, *89*, 5956–5957.

19. Hong, X.; Chen, S,; Dlott, D. D., Ultrafast mode-specific intermolecular vibrational energy transfer to liquid nitromethane. *J. Phys. Chem.*, **1995**, *99*, 9102–9109.

20. von Jena A.; Lessing, H. E., Coherent Coupling Effects in Picosecond Absorption Experiments. *Appl. Phys.* **1979**, *19*, 131–144.

21. Bertie J. E.; Lan Z., Infrared studies of Liquids XX: The intensity of the OH stretching band of liquid water revisited, and the best current values of the optical constants of $H_2O(l)$ at 25°C between 15,000 and 1 cm^{-1}. *Appl. Spectrosc.* **1996**, *50*, 1047–1057.

22. Iwata, T.; Koshoubu, J.; Jin, C.; Okubo Y., Temperature Dependence of the Mid-infrared OH Spectral Band in Liquid Water. *Appl. Spectrosc.* **1997**, *51*, 1269–1275.

23. Wyss, H. R.; Falk, M., Infrared Spectra of HDO in water and in NaCl solution. *Can. J. Chem.* **1970**, *48*, 607.

24. Nienhuys, H.-K.; van Santen, R. A.; Bakker, H. J., Orientational relaxation of liquid water as an activated process. *J. Chem. Phys.* **2000**, *112*, 8487–8494.

25. Stepanov, B. I., Interpretation of the Regularities in the Spectra of Molecules Forming the Intermolecular Hydrogen Bond by the Predissociation effect. *Nature* **1946**, *157*, 808.

26. Staib A.; Hynes, J. T., Vibrational predissociation in hydrogen-bonded OH$\cdots$O complexes via OH stretch–OO stretch energy transfer. *Chem. Phys. Lett.* **1993**, *204*, 197–205.

27. Bakker, H. J.; Woutersen, S.; Nienhuys, H.-K., Reorientational motion and hydrogen-bond stretching dynamics in liquid water. *Chem. Phys.* **2000**, *258*, 233–246.

28. Hemenger R. P.; Pearlstein, R. M., Time-dependent concentration depolarization of fluorescence. *J. Chem. Phys.* **1973**, *59*, 4064–4072.

29. Inokuti M.; Hirayama, F., Influence of energy transfer by the exchange mechanism on donor luminescence. *J. Chem. Phys.* **1965**, *43*, 1978–1989.

30. Eisenthal K. B.; Siegel, S., Influence of resonance transfer on luminescence decay. *J. Chem. Phys.* **1964**, *41*, 652–655.

31. Graener, H.; Seifert, G.; Laubereau, A., Vibrational and reorientational dynamics of water molecules in liquid matrices. *Chem. Phys.* **1993**, *175*, 193–204.
32. Szabo, A., Theory of fluorescence depolarization in macromolecules and membranes. *J. Chem. Phys.* **1984**, *81*, 150–167.
33. McGraw, R.; Madden, W. G.; Bergren, M. S.; Rice, S. A., A theoretical study of the OH stretching region of the vibrational spectrum of ice I*h*. *J. Chem. Phys.* **1978**, *69*, 3483–3496.

Chapter 15

The Relationship between the Self-Diffusivity of Supercooled and Amorphous Solid Water

R. Scott Smith, Z. Dohnálek, Greg A. Kimmel, K. P. Stevenson, and Bruce D. Kay[*]

Environmental Molecular Sciences Laboratory, Pacific Northwest National Laboratory, 3335 Q Avenue, Mail Stop K8–88, Richland, WA 99352

We summarize the existing experimental data for the self-diffusivity of supercooled liquid water and review two proposals for the temperature dependence of these data. These data are compared to the recently published measurements of the self-diffusivity of amorphous solid water. We discuss the implications of these data regarding the continuity between ASW and supercooled liquid data. The advantages and limitations of using nanoscale thin films to measure the physical properties of metastable amorphous materials are discussed.

Introduction

The properties of supercooled liquid water have been the subject of much experimental and theoretical research and several excellent reviews of this work are available. The experimental properties have been summarized in a review article by Angell (*1*). The theoretical interpretation of the experimental data are thoroughly discussed by Debenedetti (*2*). More recently, Mishima and Stanley discuss some of the current puzzles and outstanding issues regarding liquid, supercooled liquid, and glassy water (*3*).

One issue is the relationship between amorphous forms of solid water and liquid water. Water forms an amorphous solid when vapor deposited on cold (<140 K) substrates (*4*). This amorph is known as Amorphous Solid Water (ASW). (Other amorphous solids of water have also been formed by rapid quenching of liquid water (*5*) and by high pressure amorphization of crystalline ice (*6,7*), however, the similarities and differences between these amorphs and ASW are beyond the scope of this paper). The question is whether or not the melt of ASW when heated above its glass transition temperature is thermodynamically continuous with normal supercooled liquid water. There are several theoretical proposals for and against continuity (*2*), however unambiguous resolution has not been achieved.

In this paper, we discuss the continuity issue in terms of a single physical property — the self-diffusivity. We review the existing supercooled liquid and ASW diffusivity data and discuss arguments for and against continuity. Finally, we discuss the use of nanoscale thin films to determine the physical properties of amorphous solids. Part of this paper, namely the data and discusion related to Figure 5, has been reported previously (*8,9*).

Supercooled Liquid Water Diffusivity

The self-diffusivity of supercooled liquid water has been measured by Gillen et al. (*10*), by Prielmeier et al. (*11*), and more recently by Price et al. (*12*). These data, along with the liquid water diffusivity data compiled by Weingärtner (*13*), are plotted in Figure 1. The Prielmeier et al. (squares), Price et al. (diamonds), and Weingärtner (circles) data are remarkably self-consistent and show a smooth temperature dependence between the liquid and supercooled liquid diffusivity. The Gillen et al. data (plus symbols) exhibit a qualitatively similar temperature dependence but are lower in magnitude by ~5% at 300 K and by ~20% at 242 K. For clarity and because the Prielmeier et al., Price et al., and Weingärtner data are in such good agreement we will use these data in our analysis. We will refer to this data set as the liquid and supercooled liquid diffusivity data.

An Arrhenius plot of the liquid and supercooled liquid self-diffusivites is shown in Figure 2. The solid line is a fit to the liquid diffusivity data and this fit is extrapolated into the supercooled region. The supercooled liquid data deviate from this fit and show increasing non-linearity at low temperatures. This non-Arrhenius is known to occur in many glass-forming liquids as they are supercooled below their freezing point and has been observed previously for water (*1,10-12,14*). Two equations have been used to fit the temperature dependence of this data.

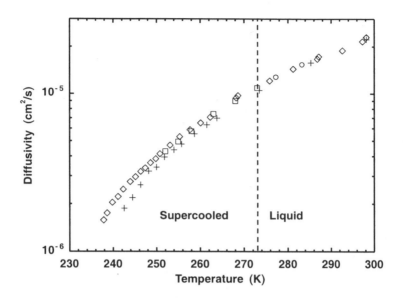

Figure 1. Liquid and supercooled liquid self-diffusivity data for water.

One approach is to use the Vogel-Fulcher-Tamman (VFT) equation,

$$D(T) = D_o \, exp\{-B/(T-T_o)\} \tag{1}$$

where T is temperature, and T_o, D_o, and B are fit parameters. This equation is often used to parameterize the viscosity of supercooled liquids and is highly accurate over several orders of magnitude for many substances. In Eqn 1, T_o is the temperature where the diffusivity descends to zero (conversely, where the viscosity goes to infinity). The magnitude of B determines the steepness of this descent. Although originally used as an empirical formula (*15*), the B and T_o parameters are thought to have thermodynamic significance and are used to used to classify supercooled liquids as either strong or fragile (*2,16*).

The other approach is to use a power law equation of the form,

$$D(T) = D_o \, T^{1/2} \, (T/T_s -1)^{\gamma} \tag{2}$$

where T is temperature, and T_s, D_o, and γ are fit parameters. This equation predicts a divergence as T approaches T_s. Many of the thermodynamic and dynamic properties of supercooled liquid water can be fit to a power law type equation with a consistent value for $T_s \approx 228$ K (*1,2,17-20*). This observation along with other evidence has led to the proposal of a singularity at 228 K in the water phase diagram. Several explanations have been proposed to address this issue (*1,14,17-29*) and the arguements are summarized very well by Debenedetti (*2*). Below we illustrate the issue by showing fits to the temperature dependence of the self-diffusivity.

Fits of the liquid and supercooled liquid data to the VFT (solid line) and power law (dashed line) equations are shown in Figure 3. The parameters for the VFT fit are $D_o = 4 \pm 1 \times 10^{-4}$ cm^2/s, E = 370 ± 80 K, and $T_o = 170 \pm 10$ K. The power law fit parameters are $D_o = 8 \pm 4 \times 10^{-6}$ cm^2/s, $\gamma = 1.86 \pm 0.4$, and $T_s = 216.4 \pm 10$ K (similar parameters were obtained by Price et al. (*12*) and by Prielmeier et al. (*14*)). Both equations fit the data well with the power law equation fitting slightly better at temperatures below 240 K. The slightly better fit of the power law equation has been reported previously and has been used as evidence in support of a singularity at ~228 K proposal (*1,11,12,14,20*). Clearly, the existing supercooled diffusivity data do not go to low enough temperature to unambiguously predict the diffusivity below 240 K. The inability to supercool liquid water to temperatures approaching the apparent divergence at 228 K (*1,2*) has made unambiguous experimental resolution of the singularity question difficult.

Amorphous Solid Water Diffusivity

Water is known to form an amorphous solid when vapor deposited on a cold substrate (T < 140 K)(*4*). This material, referred to as amorphous solid water (ASW), is reported to have a calorimetric glass transition between 124 K and 136 K (*5,30*). The glass transition, T_g, is the temperature at which the structural relaxation time of the supercooled liquid becomes comparable to an experimental time scale, e.g. 100 s. Figure 4 shows a model phase diagram (solid line) for supercooled liquid, glass, and crystal formation. If a liquid is cooled below its melting temperature, T_m, at a rate that is faster than the crystallization rate, then a material can sometimes exist as a supercooled liquid that is thermodynamically metastable with respect to the crystalline phase. Upon further cooling the structural relaxation timescale increases until the supercooled liquid no longer behaves ergodically on an experimental timescale. The temperature where this occurs is T_g. Below T_g the relaxation timescale increases rapidly (the viscosity diverges) and the liquid's structure is "frozen" and the material behaves less like a liquid and more like a solid i.e. an amorphous solid or glass.

In principle then, one could create a metastable deeply supercooled liquid by heating a glass above its T_g. Such an experiment is illustrated by the dashed line in Figure 4. When heated above T_g, the material would be a supercooled liquid that could at some point undergo an irreversible transformation to the

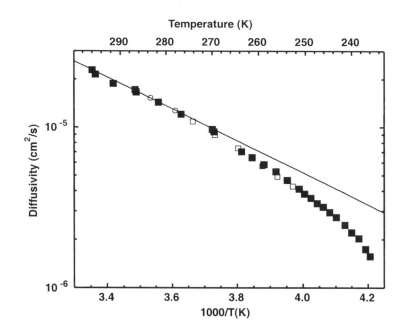

Figure 2. Arrhenius plot of the liquid and supercooled self-diffusivity data.

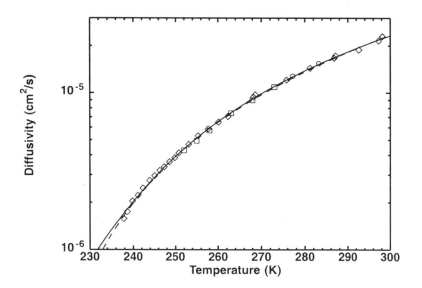

Figure 3. VFT and Power Law fits to the liquid and supercooled self-diffusivity.

crystalline state. The lifetime of the metastable liquid prior to crystallization would determine whether or not one could measure its diffusivity. For example, at T_g a material would have a diffusivity of $\sim 10^{-20}$ cm^2/s (a value commonly used to define T_g and corresponds to a viscosity of $\sim 10^{13}$ poise) and this would result in motion of ~ 1 molecular diameter (~ 3Å for water) in about 100 s. Such a diffusivity would be nearly impossible to observe with a macroscopic sample — a 1 cm thick film would require $\sim 10^{12}$ years to completely mix!

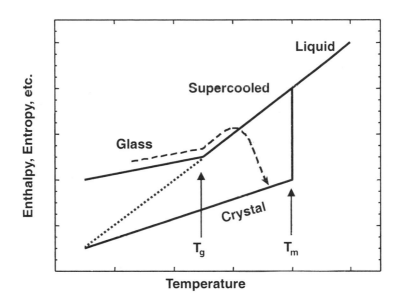

Figure 4. Schematic of a model liquid, supercooled, and glass phase diagram.

In recent experiments, our group has used nanoscale thin films to overcome the problem of observing extremely small diffusion lengths on a reasonable laboratory time scale (~ 10-100 s) to determine the self-diffusivity of ASW upon heating above its T_g (*8,9,31*). In these experiments nanoscale films of varying isotopic composition are created by sequential dosing of the isotopic vapor at low temperature (~ 85 K). After ASW film growth, the sample temperature was raised above the glass transition ($T_g \sim 140$ K), through the crystallization region (~ 155-160 K), and to a temperature where the ice film had completely desorbed (~ 170-180 K). The desorption spectra during this temperature ramp reveal the crystallization kinetics of the ASW films and also the extent of intermixing between the $H_2^{16}O$ and $H_2^{18}O$ layered interfaces. The temperature dependent diffusivity is quantified using a mathematical model that couples a mean-field description of the desorption/crystallization kinetics to a one-dimensional representation of the diffusive transport between layers (*8,9*). The diffusion rate

204

is dependent on the phase of the material (amorphous or crystalline) and the diffusion rate is treated as a linear combination of the amorphous and crystalline diffusivity weighted by their respective mole fractions. Upon crystallization the diffusive motion is effectively "frozen" out. The results were independent of substrate. Details on the experimental technique, analysis, and results are discussed elsewhere (*8,9*), so here we present only the ASW diffusivity results.

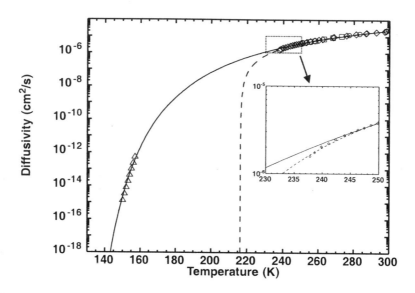

Figure 5. VFT fit to the liquid, supercooled, and ASW self-diffusivity data. (Reproduced with permission from reference 8. Copyright 1999 Macmillan Magazines Limited.)

A plot of the ASW diffusivity, along with the supercooled liquid, and liquid water diffusivities is shown in Figure 5. The individual ASW diffusivity points (open triangles) from 150-157 K, the temperature range where we observe significant ASW diffusion prior to crystallization are calculated using the Arrhenius equation used to fit the data in the model simulations ($D_o = 2.9\times10^{43}$ cm^2/s and E = 170 kJ/mole). The solid line in Fig. 5 is the result of fitting the ASW and the liquid/supercooled liquid diffusivity data to the VFT equation (Eqn. 1). The VFT fit parameters are $D_o = 2.8 \pm 2\times10^{-3}$ cm^2/s, E = 864 ± 100 K, and T_o = 119 ± 3 K (*8,9*). Also shown in Fig. 5 is a fit of the power law equation to the liquid/supercooled liquid diffusivity from Figure 3. The VFT equation appears to fit the liquid, supercooled liquid, and the ASW data. On an expanded scale (Fig. 5 inset) the VFT fit deviates from the data below 240 K by about 25% and the power law gives a better fit. On the other hand, the power law equation with a singularity near 220 K fits only the liquid and supercooled liquid data and not the ASW data. This is not a surprise since the ASW data were not included in the fit. Obviously, any equation with a singularity above the temperature of the ASW data cannot simultaneously fit both the ASW and liquid diffusivity data.

The Continuity Question

The open question is whether the melt of ASW is an extension of the supercooled liquid. That is, whether there is continuity between the normal (1 atm pressure) supercooled liquid diffusivity and the liquid-like diffusivity near 150 K. The VFT and power law equations give two distinct answers to this question. On one hand, if one assumes that the data can be connected, then the VFT equation describes the entire set of data spanning a range of 10^{10} reasonably well (within ± 25% deviation). This interpretation of the data does not require either a singularity at 228 K or a new distinct phase of liquid water. An extrapolation of the VFT fit to 10^{-20} cm^2/s (a common definition of T_g) predicts a T_g of about 141 K (2,16) which compares favorably to the experimentally determined calorimetric glass transition observed between 124 K and 136 K (5,30). On the other hand, the results of Figure 5 (inset) confirm the accuracy of the power law fit to the supercooled liquid data and suggest that the melt of ASW is not connected to normal supercooled liquid water. This interpretation predicts a singularity at ~220 K and would require the existence of a new phase of liquid water to account for the ASW diffusivity.

Both of these interpretations require assumptions about the diffusivity in temperature regions where data do not presently exist. The VFT fit requires interpolation between ~240 K and ~160 K. The missing diffusivity data in this 80 K wide region could display a temperature dependence that is consistent with both the ASW data and the power law. The power law fit requires an extrapolation of about 18 K between the lowest experimental data point (~238 K) and the singularity at 220 K. In general, data much closer to the proposed singularity are needed to confirm its existence. Unfortunately experiments in the 160-240 K temperature region have not been possible due to the rapid crystallization of supercooled liquid water below 240 K and of ASW above 160 K.

Other Experimental Evidence

While the diffusivity data alone do not provide unambiguous support for continuity, the diffusivity temperature dependence is consistent with other experimental data. As discussed above, an extrapolation of the VFT fit to 10^{-20} cm^2/s predicts a T_g of about 141 K. The rapid decrease in diffusivity upon cooling towards the glass transition temperature is consistent with a fragile material (2,16). A highly temperature dependent diffusivity at low temperatures is also consistent with the observations of Fisher and Devlin (32). They report no translational diffusion more than a length scale of a few molecular diameters at 125 K within 13.5 hours (*32*). Our VFT extrapolation to 125 K would predict that there would be no observable translational diffusion on the timescale of those experiments. Later in Figure 8, we discuss reasons why there should be no diffusion below 140 K.

Other experimental evidence that supports the continuity interpretation comes from mechanical deformation measurements of low density amorphous (LDA) (*33*). LDA is a water amorph formed by pressure induced collapse of crystalline ice and subsequent annealing. While this amorph may have properties slightly different than those of ASW, it has the same density and it is believed to behave analogously above its T_g (~130 K). The deformation studies show that LDA behaves like a viscous liquid when heated above its T_g (*33*). While the quantitative determination of the viscosity was not possible, the liquid-like LDA results were dramatically different than those obtained for cubic ice and were analogous to the results obtained for other glass forming materials.

Additional experimental evidence comes from the temperature dependence of the dielectric relaxation data from Bertolini et al. (*34*) (252-305 K) and Johari (*35*) (163-174 K). The reciprocal of the dielectric relaxation time (used to facilitate a comparison with the diffusivity) and the diffusivity have a very similar temperature dependence and have similar VFT parameters (*9*). This is expected since both processes are related to the same microscopic molecular motions. It is possible to estimate the magnitude of the diffusivity from the dielectric relaxation time using the relationship $D \sim a^2/\tau$, where a is a characteristic length and τ is the relaxation time. At 163 K, Johari (*35*) finds that τ is ~10^3 seconds. Assuming a hopping length characteristic of a molecular diameter, ~3Å. yields a diffusivity of ~10^{-12} cm^2/s. This value is close to the values we determine in our diffusivity experiments. Furthermore, the strong temperature dependence of the low temperature dielectric relaxation is in accord with the temperature dependence we obtain for the diffusivity.

The proposal of a high temperature singularity is based primarily on the power law temperature dependence of many of the properties of supercooled liquid water (*1,2*). This proposal was also supported by the inability to supercool liquid water below ~233 K (*1,2*). It was suggested that this may be an indication of an underlying thermodynamic property rather than a kinetic limitation (*1*). There is, however, more recent evidence that shows that water can be supercooled below 233 K by using cluster droplet (*36*) and liquid jet expansions (*37*). Both of these experiments rely on rapid evaporative cooling to obtain cooling rates between 10^5 K/s (*37*) and 10^7 K/s (*36*). The water cluster experiments use electron diffraction to show that water clusters have a liquid-like diffraction pattern down to about 200 K (*36*) prior to freezing into a crystalline solid. In the liquid jet experiments, a thin (5 μm to 50 μm) stream of liquid water is expanded into a vacuum and the liquid is observed to reach a temperature of about 210 K (*37*). The temperature is determined by measuring the velocity distribution of evaporating water molecules from a liquid water jet. These combined results suggest that the inability to supercool bulk liquid water samples below 233 K (*1*) arises from kinetic effects rather than the existence of an underlying thermodynamic singularity.

Diffusivity and Viscosity

The diffusivity and the viscosity are inversely related functions. In liquids this relationship is often well described by the Stokes-Einstein equation, $D = kT / 6\pi a\eta$, where η is the viscosity and a is the molecular radius. In supercooled liquids this relationship can breakdown at low temperatures.

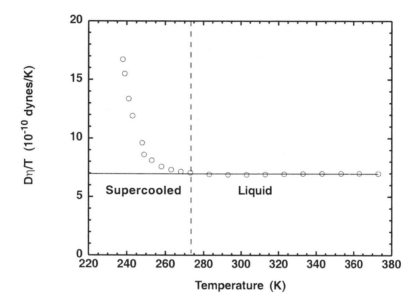

Figure 6. Plot of $D\eta/T$ quantity for liquid and supercooled water.

In Figure 6 the quantity $D\eta/T$ (open circles) is plotted versus temperature where the D is from the VFT curve in Figure 5 and η is from reference (*1*). According to Stokes-Einstein this quantity should be a constant. In the liquid region the quantity is constant (solid line), but at lower temperatures in the supercooled region there is a positive deviation from this constant value. The positive deviation is the result of the more rapid increase in the viscosity compared to the decrease in the diffusivity. While the exact physical reason for the breakdown of Stokes-Einstein is not known, it does suggest that the determination of the supercooled water diffusivity from viscosity experiments (and vice versa) is not possible.

Lifetime of Metastable Amorphous Thin Films

The temperature region over which we measure significant diffusivity in the ASW deposit is relatively narrow (~150–157 K). The reason for this is due to two factors. The first is the temperature dependence of the ASW crystallization. Figure 7 shows a plot of the ASW crystallization time as a function of temperature (dashed line). This time is calculated using the ASW crystallization kinetics and is the time for crystallization at a given isothermal temperature (8,31,38,39). The isothermal crystallization time varies from thousands of seconds at 140 K to a few seconds near 160 K. This time sets the period during which diffusion can occur since after crystallization the diffusivity in crystalline ice is extremely low and motion is essentially "frozen". The second factor is the temperature dependence of the diffusivity itself. The diffusivity is plotted as a solid line in Figure 7. The diffusivity is highly activated in this temperature region and the diffusivity ranges from ~ 10^{-18} at 142 K to more than 10^{-12} near 160 K.

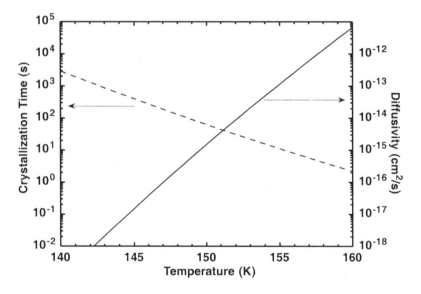

Figure 7. Plot of the crystallization time and diffusivity for ASW.

The combination of these two factors result in a narrow temperature region over which observable diffusion can be measured. At low temperatures the crystallization time is long, but the diffusivity is small. Conversely at higher temperatures the diffusivity is large but the crystallization time is short. As a result, in both cases the overall diffusion length is limited. Figure 8 shows a plot of the expected diffusion length at a given temperature. This is calculated using the equation $L = (D\tau)^{1/2}$, where the crystallization time, τ, and the diffusivity D, are taken from the data in Figure 7. The vertical arrows indicate the calculated diffusion lengths where the crystallization time is 1000, 100, and 10 seconds. For example, at ~146 K, the isothermal crystallization time is about 1000 s and during this time a water molecule would diffuse ~15Å. On the other hand, at ~156 K, the isothermal crystallization is about 10 s and during this time a molecule would diffuse about 150 Å. Also notice that below 140 K there should be no diffusion at all (regardless of time) which is consistent with the results of Fisher and Devlin (32).

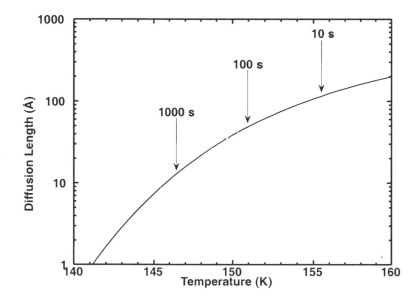

Figure 8. Plot of the calculated diffusion length versus temperature for ASW.

Because crystallization and diffusion are kinetic processes that have different activation energies the diffusion length will be dependent on the rate of the temperature ramp. Changing the temperature ramp rate would definitely change the temperature at which a sample would completely crystallize and thereby effect the diffusion length in that sample. This effect could be exploited to test the activation energies used in our simulations and to extend the temperature range where diffusion is observed. These studies will be the subject of future research.

Summary

We have used nanoscale thin films of vapor deposited water isotopes to determine the self-diffusivity of ASW upon heating above its T_g. The diffusivity in the 150 to 160 K range is highly activated (170 ± 40 kJ) and near 160 K the diffusivity of ASW reaches $\sim 10^{-12}$ cm^2/sec, a value that is roughly 10^7 times smaller than the diffusivity of normal liquid water at room temperature. Such a diffusivity would be nearly impossible to observe with a macroscopic sample — a 1 cm thick film would require $\sim 10^5$ years to completely mix! The use of nanoscale films enables these small diffusivities to be determined quantitatively. Collectively, these results combined with our previous studies addressing the consistency of a thermodynamic continuity for the free energy (*40*), suggest that at the calorimetric glass transition temperature of 136 K the amorphous solid melts into a deeply supercooled metastable extension of normal liquid water prior to crystallizing near 160 K. This interpretation does not require the existence of a temperature singularity near 228 K at low pressure (<0.1 MPa) (*1,17*), and is consistent with but not proof of the existence of a second critical point near 220 K and 0.1 GPa (*7,22-26*). While our new ASW diffusivity data provide support for a continuity between ASW and liquid water at low pressure, an unambiguous resolution of the continuity conundrum must await further experiments in the unexplored temperature region from 160 to 240 K.

This work was supported by the U. S. Department of Energy Office of Basic Energy Sciences, Chemical Sciences Division. Pacific Northwest National Laboratory is operated for the U. S. Department of Energy by Battelle under contract DE-AC06-76RLO 1830.

References

1. Angell, C. A. In Water: A Comprehensive Treatise, Volume 7; 1st ed.; Franks, F., Ed.; Plenum Press: New York, 1982; Vol. 7, pp 1.
2. Debenedetti, P. G. Metastable Liquids: Concepts and Principles; Princeton University Press, 1996.
3. Mishima, O.; Stanley, H. E. Nature **1998**, 396, 329.
4. Sceats, M. G.; Rice, S. A. In Water: A Comprehensive Treatise, Volume 7; 1st ed.; Franks, F., Ed.; Plenum Press: New York, 1982; Vol. 7, pp 83.
5. Johari, G. P.; Hallbrucker, A.; Mayer, E. Nature **1987**, 330, 552.
6. Mishima, O.; Calvert, L. D.; Whalley, E. Nature **1984**, 310, 393.
7. Mishima, O.; Stanley, H. E. Nature **1998**, 392, 164.
8. Smith, R. S.; Kay, B. D. Nature **1999**, 398, 788.
9. Smith, R. S.; Dohnalek, Z.; Kimmel, G. A.; Stevenson, K. P.; Kay, B. D. Chem. Phys. **2000**, 258, 291.
10. Gillen, K. T.; Douglass, D. C.; Hoch, M. J. R. J. Chem. Phys. **1972**, 57, 5117.

11. Prielmeier, F. X.; Lang, E. W.; Speedy, R. J.; Lüdemann, H.-D. Ber. Bunsenges. Phys. Chem. **1988**, 92, 1111.
12. Price, W. S.; Ide, H.; Arata, Y. J. Phys. Chem. A **1999**, 103, 448.
13. Weingärtner, H. Z. Phys. Chem. **1982**, 132, 129.
14. Prielmeier, F. X.; Lang, E. W.; Speedy, R. J.; Ludemann, H.-D. Phys. Rev. Lett. **1987**, 59, 1128.
15. Scherer, G. W. J. Am. Ceram. Soc. **1992**, 75, 1060.
16. Angell, C. A. Science **1995**, 267, 1924.
17. Speedy, R. J.; Angell, C. A. J. Chem. Phys. **1976**, 65, 851.
18. Speedy, R. J. J. Chem. Phys. **1982**, 86, 982.
19. Speedy, R. J. J. Phys. Chem. **1982**, 86, 3002.
20. Angell, C. A. Ann. Rev. Phys. Chem. **1983**, 34, 593.
21. Speedy, R. J. J. Phys. Chem. **1987**, 91, 3354.
22. Poole, P. H.; Sciortino, F.; Essman, U.; Stanley, H. E. Nature **1992**, 360, 324.
23. Poole, P. H.; Essman, U.; Sciortino, F.; Stanley, H. E. Phys. Rev. E **1993**, 48, 4605.
24. Poole, P. H.; Sciortino, F.; Essman, U.; Stanley, H. E. Phys. Rev. E **1993**, 48, 3799.
25. Poole, P. H.; Sciortino, F.; Grande, T.; Stanley, H. E.; Angell, C. A. Phys. Rev. Lett. **1994**, 73, 1632.
26. Stanley, H. E.; Angell, C. A.; Essmann, U.; Hemmati, M.; Poole, P. H.; Sciortino, F. Physica A **1994**, 205, 122.
27. Stanley, H. E.; Teixeira, J. J. Chem. Phys. **1980**, 73, 3404.
28. Xie, Y.; K. F. Ludwig, J.; Morales, G.; Hare, D. E.; Sorensen, C. M. Phys. Rev. Lett. **1993**, 71, 2050.
29. Borick, S. S.; Debenedetti, P. G.; Sastry, S. J. Phys. Chem. **1995**, 99, 3781.
30. Handa, Y. P.; Klug, D. D. J. Phys. Chem. **1988**, 92, 3323.
31. Smith, R. S.; Huang, C.; Kay, B. D. J. Phys. Chem. B **1997**, 101, 6123.
32. Fisher, M.; Devlin, J. P. J. Phys. Chem. **1995**, 99, 11584.
33. Johari, G. P. J. Phys. Chem. B **1998**, 102, 4711.
34. Bertolini, D.; Cassettari, M.; Salvetti, G. J. Chem. Phys. **1982**, 76, 3285.
35. Johari, G. P. J. Chem. Phys. **1996**, 105, 7079.
36. Bartell, L. S.; Huang, J. J. Phys. Chem. **1994**, 98, 7455.
37. Faubel, M.; Schlemmer, S.; Toennies, J. P. Z. Phys. D **1988**, 10, 269.
38. Smith, R. S.; Huang, C.; Wong, E. K. L.; Kay, B. D. Surf. Sci. Lett. **1996**, 367, L13.
39. Dohnálek, Z.; Ciolli, R. L.; Kimmel, G. A.; Stevenson, K. P.; Smith, R. S.; Kay, B. D. J. Chem. Phys. **1999**, 110, 5489.
40. Speedy, R. J.; Debenedetti, P. G.; Smith, R. S.; Huang, C.; Kay, B. D. J. Chem. Phys. **1996**, 240.

Metastable Liquids

Chapter 16

Spatially Heterogeneous Dynamics in Liquids Near Their Glass Transition

S. C. Glotzer[1], Y. Gebremichael[1], N. Lacevic[1], T. B. Schrøder[2], and F. W. Starr[3]

[1]Departments of Chemical Engineering and Materials Science and Engineering, University of Michigan, 2300 Hayward Street, Ann Arbor, MI 48109
[2]IMFUFA, Roskilde University, DK–4000 Roskilde, Denmark
[3]Polymers Division, National Institute of Standards and Technology, Gaithersburg, MD 20899

Liquids near their glass transition are dynamically heterogeneous, and experiments and simulations have uncovered much about the rich and complex nature of this heterogeneity. This paper highlights key results from our computer simulation studies of spatially heterogeneous dynamics (SHD) in supercooled liquids near their glass transtion. We speculate on the relationship between SHD and other phenomena ubiquitous to glasses and their liquids, and outline several outstanding questions in this field.

Computer simulations and experiments have revealed a great deal about the complex dynamics inherent to liquids approaching their glass transition. While much remains to be understood, it is fair to say that the existence of spatially heterogeneous dynamics (SHD) in these fluids is now firmly established. In this short paper, we provide a perspective that attempts to synthesize the insights gained from simulation studies of this phenomenon, and enumerate some of the many questions that remain to be addressed. As there exist several recent review and overview papers on the topic of dynamical heterogeneity [1-4], the present paper is not meant to be a comprehensive discussion; instead, we aim to provide a brief introduction to SHD, and describe our work relating to the discovery and investigation of dynamical, ordered structures within otherwise disordered, glass-forming liquids [3, 5-13].

It has long been known that despite the similarity in structure of a liquid and its glass, relaxation times, diffusivities and viscosities change by up to 14 orders of magnitude as a liquid is cooled through its glass transition. Why the dynamics can change so dramatically while static structure seemingly changes so little has been a longstanding, open question in the field of glass research. Clearly, molecular motion becomes increasingly difficult as the temperature or specific volume decreases, but how precisely does the motion change, and why does this change lead to such long relaxation times? Mode coupling theory (MCT) provides a partial answer to this question for the initial approach to the glass transition; that is, for temperatures where diffusivities are still orders of magnitude higher than at T_g [14]. The theory makes specific predictions for the slowing down of fluids from their equilibrium structure, which hold asymptotically near a crossover temperature T_{mct}, which is typically 1.1 to 1.5 times higher than T_g. In hard-sphere, colloidal suspensions, where $T_c \approx T_g$, MCT is able to describe the dynamics closer to T_g. However, the theory in its current form does not provide spatial information revealing how the dynamics change *locally* in a fluid as it is cooled, nor does it specify how molecular motion occurs.

Many clues abound that point to a dramatic change in the mechanism of molecular motion in cold, dense liquids. At sufficiently high temperatures and/or low densities, liquids (including atomic, molecular, polymeric or colloidal liquids) exhibit diffusive behavior at long times, and inertial behavior (except for colloids) at very short times. As liquids are cooled or their density increased, their dynamical behavior changes, and the "particles" comprising the fluid (atoms in, e.g., a bulk metallic glass, molecules in, e.g., water or glycerol, polymer segments in a melt or blend, or colloidal particles in a suspension) become temporarily trapped on intermediate time scales in "cages" formed by neighboring particles. This caging behavior is evidenced by the appearance of a plateau in the mean square displacement and intermediate scattering function, $F(q,t)$, which measures the decay of density fluctuations and can be obtained directly by neutron spin echo experiments or through a Fourier transform of the dynamic structure factor $S(q,\omega)$. At high temperature T, $F(q,t)$ decays exponentially in time, reflecting the Gaussian nature of the distribution of particle displacements. At lower T and or larger density ρ, it develops a plateau at intermediate times, and at longer times decays as a stretched exponential with a characteristic time τ_α that grows rapidly with decreasing T [15]. The plateau indicates that, on that time scale, the average motion of the particles is limited to a small fraction of its diameter. As the liquid is further cooled or densified, the plateau extends to longer and longer times, but provided the liquid is above (i.e. on the liquid side of) its glass transition, particles eventually break out of their cages and exhibit diffusive motion at late times.

The onset of "caging" is but one indicator of the onset of "glass-forming liquid" or "supercooled liquid" behavior. Coincident with the onset of caging is also a change in the way the liquid explores its potential energy landscape [16], and a change in the relaxation of the intermediate scattering function from exponential to stretched exponential. Below the onset of caging, the characteristic structural (or "alpha") relaxation time τ_α increases rapidly with decreasing T, exhibiting power law behavior approaching T_{mct}, and Vogel-Fulcher-Tammann behavior below T_{mct}. Both the appearance of the plateau and the non-exponential decay indicate an increasing complexity in the fluid dynamics MCT, which is a theory based on the caging of molecules and on the collective motions associated with caging, describes well much of the behavior of F(q,t) above T_{mct} [15, 17-20].

Another dramatic indication of a change in molecular mechanism in equilibrium liquids on cooling is provided by the relationship between translational and rotational diffusivity and viscosity. At high T, these transport coefficients are related by the Stokes-Einstein (SE) and Stokes-Einstein-Debye (SED) relations. At sufficiently low T, the SE relation breaks down, and the diffusion coefficient of a probe in the fluid is higher (in some instances, by several orders of magnitude) than would be predicted by the viscosity from the SE relation [2, 21-23]. This so-called "decoupling" of transport coefficients demonstrates a breakdown of continuum hydrodynamics, and an increasing complexity in the fluid dynamics [22, 24, 25].

Through a combination of multidimensional nuclear magnetic resonance (NMR) experiments, optical bleaching, flourescence and solvation spectroscopy experiments, dielectric hole burning experiments, atomic force microscopy (AFM) probed polarization noise, triplet-state solvation dynamics, and computer simulations, there is an abundance of evidence that both stretched exponential relaxation and the decoupling of transport coefficients can be rationalized if glass-forming liquids are *spatially heterogeneous in their dynamics* (SHD) [2, 4, 25-30]. Indeed, we now believe that as the glass transition approaches, particle (or molecule) motion becomes strongly correlated over increasingly larger distances. This leads to an emergence of increasing dynamical order in the fluid, and to the appearance of large-scale fluctuations in the local molecular mobilities.

Computer simulation is helping us to investigate these spatiotemporal fluctuations in the local mobilities, as well as to elucidate the precipitating events that we believe initiate high mobility regions. In particular, simulation has allowed us to make specific predictions regarding the nature of these regions [3, 5-10], including their size, shape, and dynamical character ---- predictions that have now been confirmed in experiments [31-34].

Two of the most often used models for studying the dynamics of glass-forming liquids and polymers are the LJ binary mixture model and bead-spring model, respectively. We have investigated correlated particle motion in molecular dynamics simulations of two different binary mixtures – the 80:20 mixture of Kob-Anderson (KA), and a 50:50 mixture, both containing 8000 particles. In both models, the interaction parameters are chosen to prevent phase separation and crystallization. We have also studied the dynamics of a polymer melt described by a bead-spring model, where again the interaction parameters are chosen to prevent crystallization. In this system, we simulate roughly 120 chains of 10 monomers each, short enough that the melt is unentangled. The details of our simulations are discussed elsewhere [3, 5-10]. In all systems, we perform simulations in the NVE or NVT ensembles at many different state points, in equilibrium above or at the mode-coupling temperature T_c. Typically, $T_c \approx 1.1$-$1.5\ T_g$ in these models. Our longest simulations span more than 10^8 MD steps, or tens to hundreds of nanoseconds per run in Argon units.

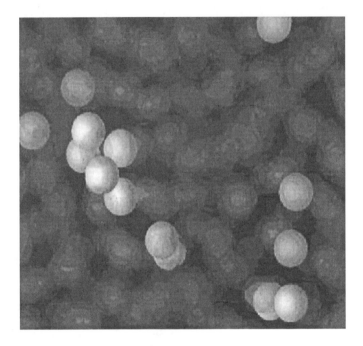

Figure 1. Highly mobile or "fluidized" regions in a simulated LJ liquid at a termperature above its glass transition. Particles escaping from their "cages" at the time of observation are lightly colored, while particles still trapped in their cages are colored darker gray and faded out. Image demonstrates the spatially heterogeneous nature of the particle dynamics.

By visualizing the raw data from the simulation and watching the motion of individual particles (monomers in the case of the melt), it is immediately apparent that at temperatures below that corresponding to the onset of caging, particle motion becomes "intermittent" [35]; in any given time interval, most of the particles are localized in cages formed by their neighbors, and only a handful (between 5 and 12%) escape from their cages at a time [5, 36]. It is also obvious from watching the simulation that below the "onset" temperature, particles escape from their cages in "groups" or "clusters", and thus molecular motion becomes spatially correlated (heterogeneous) on cooling. Thus over long times, many particles remain fixed relative to their initial position, and within this "immobile" matrix emerge "fluidized patches" within which particles move (Figure 1) [25]. It is also immediately evident from the simulation that the number of particles involved in a typical cluster grows with decreasing T, indicating the growing range of spatially heterogeneous dynamics on cooling.

This heterogeneity can be quantified using methods from percolation theory and by constructing suitable correlation functions, such as displacement-displacement correlation functions and four-point, time-dependent density correlation functions [3, 5-13]. Different statistical quantities highlight different aspects of dynamical heterogeneity; some focus on the mobile regions, and some on the immobile regions, of the fluid.

For example, spatially heterogeneous dynamics can be easily observed by comparing snapshots of a glass-forming liquid at two different times, provided the time interval between them is not so short that particle motion is ballistic (in the case of atomic, molecular or polymeric fluids), and not so long that particle motion is diffusive. Confocal microscope images of suspensions of effectively hard-sphere colloids can be used for this purpose, and the "difference" between two snapshots separated by a timescale in the plateau region indicates regions of "activity" and "inactivity"; that is, regions where particles appear not to have moved from their initial positions, and regions where rearrangement has taken place. From these snapshots, it is possible to construct a correlation function $g_4(r,t)$ and corresponding generalized susceptibility $\chi_4(t)$ related to the fluctuations in the number $Q(t)$ of "overlapping" particles; that is, particles which appear not to have moved during the time interval of the comparison [12, 37-40]. We have calculated these quantities in our simulations, and shown that the fluctuations in $Q(t)$ depend on the time interval of observation, peak near the α-relaxation time τ_α and increase rapidly with decreasing T (Figure 2) [12, 39, 40]. Correspondingly, the correlation function $g_4(r,t)$ of overlapping particles develops increasing structure, and apparently increasing range, with decreasing T [40]. Thus a correlation length associated with overlapping particles can be measured directly from $g_4(r,t)$ in simulations and in microscopy studies of colloidal suspensions. Our simulations predict that this length begins to increase below the temperature corresponding to the onset of caging, and continues to increase on cooling or densification. Theoretical calculations support the simulation predictions, and due to the mean-field nature of the calculations predict a diverging length scale at T_{mct} [38, 39].

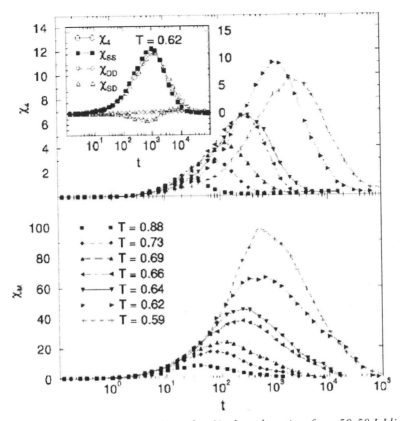

Figure 2. Susceptibilities $\chi_4(t)$ and $\chi_M(t)$ plotted vs. time for a 50:50 LJ liquid (simulation of 500 particles). $\chi_4(t)$ is similar to $\chi_M(t)$ described in the text, but is calculated with the square of the particle displacements. Note that at each T, $\chi_4(t)$ is maximum at a later time than is $\chi_4(t)$. Due to the small system size, we observe finite size effects at the lower temperatures ($\chi_4(t)$ is suppressed relative to its value in an 8000 particle simulation)[40]. Inset in top figure shows that $\chi_4(t)$ is dominated by localized particles. Reprinted with permission from [12].

The overlapping particles described above are dominated by localized particles (Figure 2 inset), but a simple replacement of Q by 1-Q in the theory essentially transforms localized regions into delocalized regions and vice versa [12]. To gain further insight into the dynamically heterogeneous nature of these liquids, and to specifically explore the high-mobility fluctuations, correlations functions of particles that incorporate the scalar or vector displacement can be used. We constructed such a correlation function whose corresponding generalized susceptibility is proportional to the fluctuations in the total system displacement U in a given time interval [8, 41]. By construction, the correlation function is most heavily weighted by particles with large displacements. The susceptibility $\chi_U(t)$ and corresponding correlation function $g_U(r,t)$ display qualitatively the same time-dependent and temperature-dependent behavior as $\chi_4(t)$ and $g_4(r,t)$, but peak and have the longest range, respectively, at a time in the late beta/early-alpha relaxation regime; i.e. at a time that precedes, and scales with T differently than, τ_α. This suggests that many high mobility fluctuations – that is, spatially correlated particle rearrangements – are necessary for structural relaxation. In both the 80:20 LJ mixture and the polymer melt, $\chi_U(t)$ is well fitted by a power law with singular temperature $T_c = T_{mct}$. Although we do not necessarily expect a true divergence of the susceptibility or correlation length at T_{mct}, the behavior of the data demonstrates the rapidity with which these regions grow. Whether these quantities cross over to slower growth below T_{mct}, become constant, or even shrink will require further investigation of substantially lower T simulations, which currently poses a substantial computational challenge because of the prohibitively long relaxation times required to properly equilibrate the liquid.

To further characterize the nature of the mobile regions of glass-forming liquids, we have performed detailed studies of the clustering of highly mobile particles (i.e., particles that, in a given time interval, exhibit the largest scalar displacement) [7, 9, 10, 36]. We find that these particles form non-compact, highly ramified clusters whose size depends on the time interval of observation (Figs. 1 and 3). This behavior is consistent with the time-dependent behavior of the susceptibilities $\chi_4(t)$ and $\chi_U(t)$. Notably, the largest clusters are observed at a characteristic time which scales with T like the MCT β-relaxation timescale τ_ε. The cluster size distribution approaches a power law as $T \rightarrow T_{mct}^+$ with an exponent near two, and the mean cluster size grows rapidly; in the polymer system, the maximum weight-average cluster size grows from roughly two to nearly 20 monomers over the range of T studied. Depending on the system, we have found both power laws and Vogel-Fulcher expressions fit the mean cluster size data well. It is important to note that in all systems studied, and at any given $\{T,\rho\}$, *a highly mobile subset of particles can be found that maximizes the mean cluster size*. Depending on T, this subset constitutes between roughly 5 and 12% of the total particles in the system [36].

Upon closer inspection, we find that within any cluster of mobile particles, smaller subsets move together coherently, following each other to form a quasi-one-dimensional "string" [6]. In the case of the 80:20 LJ mixture, these strings have an exponential length distribution, with an average size that grows slowly with decreasing T. If fitted by a power law over the relatively narrow range of temperatures simulated, the average string length would appear to diverge at a temperature near the Kauzmann temperature, significantly outside the range of our simulations [3].

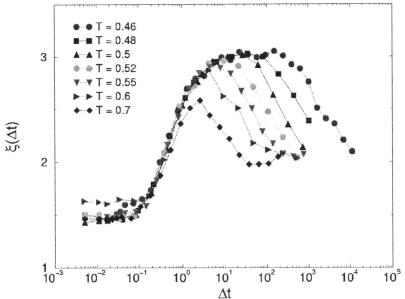

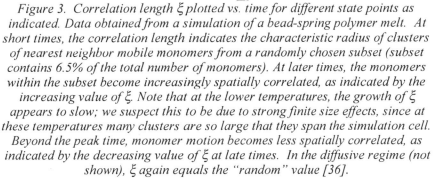

Figure 3. Correlation length ξ plotted vs. time for different state points as indicated. Data obtained from a simulation of a bead-spring polymer melt. At short times, the correlation length indicates the characteristic radius of clusters of nearest neighbor mobile monomers from a randomly chosen subset (subset contains 6.5% of the total number of monomers). At later times, the monomers within the subset become increasingly spatially correlated, as indicated by the increasing value of ξ. Note that at the lower temperatures, the growth of ξ appears to slow; we suspect this to be due to strong finite size effects, since at these temperatures many clusters are so large that they span the simulation cell. Beyond the peak time, monomer motion becomes less spatially correlated, as indicated by the decreasing value of ξ at late times. In the diffusive regime (not shown), ξ again equals the "random" value [36].

These strings likely represent the fundamental particle motions underlying the back-flow process on which mode-coupling theory is based. Additionally, they may represent the cooperative groups envisioned by Adam and Gibbs. It appears that these strings are not perfectly coherent in the T range of our simulations, in that they do not perfectly replace other particles. If they did, these string-like motions would not contribute directly to the decay of density fluctuations, since the system before and after a perfectly coherent, string-like rearrangement would be identical. Instead, the strings appear to facilitate the motion of their neighbors by perturbing the neighborhood around them. This idea is supported by studies of transitions between basins of the potential energy landscape, which at and below T_{mct} appear to be facilitated by the string-like motion of large groups of particles, which accompany many small displacements of surrounding particles [13]. Thus clusters of highly mobile particles as discussed above consist of both strings and particles whose motion is facilitated by strings. As T decreases, more and more mobile particles appear to move in strings [6].

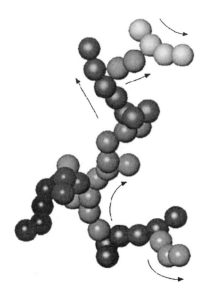

Figure 4. Schematic of a large cluster of particles in the 80:20 LJ liquid imaged from the subset of particles that exhibit the largest displacement in a time interval near the end of the plateau (caging) regime. The cluster is composed of many individual sub-groups of particles that move coherently in one-dimensional "strings", where each particle moves nearly one particle diameter in the direction of its highly mobile neighbor The actual motion within a string is relatively fast, and thus different strings within a single cluster may be "active" at different times. Reprinted with permission from [3].

Simulations clearly indicate the presence of a growing dynamical correlation length, despite the absence of a growing static correlation length. However, the local dynamics of a molecule should be related in some way to the structure of its neighborhood. Preliminary results [42] indicate a positive correlation between the Voronoi volume of a particle and its mobility, so that the larger the volume, the larger the mobility, in agreement with previous studies.

This result is consistent with previous findings that highly mobile particles tend to have a larger potential energy before becoming mobile [7], implying that the immediate neighborhood around a particle may become slightly expanded just prior to motion. These results are also consistent with recent findings of Dzugutov, et al [43] of a connection between icosahedral (non-icosahedral) ordering and low (high) mobility in a one-component glass-former. Detailed investigations of the connection between local structure and dynamics are ongoing [42].

The discovery of SHD in supercooled liquids has forced the re-examination of theories of these systems, such as mode coupling theory [44]. That MCT can describe several key aspects of the bulk dynamics of liquids in the same temperature and density regime where liquids become dynamically heterogeneous, despite the fact that MCT neither predicts nor specifically considers SHD, may not be a coincidence. In the current formulation of MCT, a rather severe approximation is made whereby fluctuating forces obey Gaussian statistics, and higher order functions are expressed in terms of two-point density functions [44]. As we have shown, a description of SHD in terms of density fluctuations requires the consideration of higher order density functions than presently considered by MCT. Thus, by construction, the present version of the theory cannot provide information on local spatial fluctuations in dynamics. However, we have shown that certain quantities associated with SHD can be used to "detect" T_{mct}, and the time scales characterizing correlations within mobile and immobile regions scale like the beta- and alpha-relaxation times, respectively. These facts, and the fact that T_{mct} may, in fact, be a crossover temperature below which the dynamics of liquids is "landscape-dominated", suggests that while MCT is too "coarse" to describe SHD, it is sufficient for describing certain bulk quantities connected to SHD. As a possible analogy, consider the phase transition in the classic Ising ferromagnet. Consideration of the magnetization alone (a one-point function) is sufficient to detect a transition from a paramagnetic state (M=0) to a ferromagnetic state (M≠0), but to detect spatial correlations in the spins, the characteristic length scale associated with spin-spin correlations, and the divergence of this length at the critical point, it is necessary to consider a two-point function, namely, the spin-spin correlation function. Similarly here, while a two-point time dependent function of densities is sufficient to detect a "transition" from one type of liquid behavior to another, the consideration of a four-point, spatiotemporal function is necessary to describe "the rest of the story". Ongoing theoretical developments, supplemented with critical tests by simulation and experiment, promise to shed light on these ideas.

In summary, with some help from simulation, the spatially heterogeneous nature of the dynamics of glass-forming liquids is now clearly established, but many open questions remain, and many connections remain to be made between the different phenomena that accompany the glass transition. Presumably, the increasing dynamical correlation length – which corresponds to the distance over which particle motion is correlated – is responsible for the increasing relaxation times that plague liquids as they are cooled. Indeed, the correlation length associated with "overlapping" regions of the fluid appears to be largest at a time that scales with T like the structural α-relaxation time. As remarked above, can a revised mode coupling theory, which includes higher order density correlations and goes beyond the usual Gaussian approximation, predict our results? How universal is spatially heterogeneous dynamics; for example, does the presence or details of SHD depend on the fragility of the glass-former? How do the length scales observed in the simulation of simple model liquids map onto real systems, or more realistic (i.e. atomistic) simulations? How are the mobility fluctuations observed in simulations above T_{mct} related to those observed in experiments near T_g? The growing mobility fluctuations in glass-forming liquids are reminiscent in some respects of growing density fluctuations in liquids near critical points; how useful is such an analogy? Are string-like rearrangements related to the Boson peak? What happens to spatially heterogeneous regions in liquids quenched to the glass state? Do high mobility regions become trapped in the glass, and govern aging, creep, shear banding, etc? What about supercooled liquids prior to nucleation and growth of the crystal phase? Are their dynamics spatially heterogeneous, and, if so, what are the ramifications for theories of nucleation? Are force chains present in glass-forming liquids with attractive interactions, and how are these structures related, if at all, to SHD? Can we use information on cooperative or correlated dynamics to construct acceleration algorithms for glassy systems, especially at low T where dynamics may be dominated by rare events? Simulation will continue to provide information necessary to answer some of these questions, and will continue to make important contributions to the general understanding of liquids and the glass transition.

Acknowledgments

SCG is especially grateful to the co-authors and to M. Aichele, P. Allegrini, J. Baschnagel, C. Bennemann, C. Donati, J.F. Douglas, S. Franz, W. Kob, V. Novikov, G. Parisi, S.J. Plimpton, P.H. Poole, and S. Sastry for their collaboration on various aspects of the work described here, and to M. Fuchs, W. Gotze, D. Reichmann, J. Schofield and many others for interesting discussions.

References

1.   Bohmer, R., *Nanoscale heterogeneity of glass-forming liquids: experimental advances.* Current Opinion in Solid State & Materials Science, 1998. **3**(4): p. 378-385.
2.   Ediger, M.D., *Spatially heterogeneous dynamics in supercooled liquids.* Annual Review of Physical Chemistry, 2000. **51**: p. 99-128.
3.   Glotzer, S.C., *Spatially heterogeneous dynamics in liquids: insights from simulation.* Journal of Non-Crystalline Solids, 2000. **274**(1-3): p. 342-355.
4.   Sillescu, H., *Heterogeneity at the glass transition: a review.* Journal of Non-Crystalline Solids, 1999. **243**(2-3): p. 81-108.

5. Kob, W., C. Donati, S.J. Plimpton, P.H. Poole, and S.C. Glotzer, *Dynamical heterogeneities in a supercooled Lennard-Jones liquid.* Physical Review Letters, 1997. **79**(15): p. 2827-2830.

6. Donati, C., J.F. Douglas, W. Kob, S.J. Plimpton, P.H. Poole, and S.C. Glotzer, *Stringlike cooperative motion in a supercooled liquid.* Physical Review Letters, 1998. **80**(11): p. 2338-2341.

7. Donati, C., S.C. Glotzer, P.H. Poole, W. Kob, and S.J. Plimpton, *Spatial correlations of mobility and immobility in a glass- forming Lennard-Jones liquid.* Physical Review E, 1999. **60**(3): p. 3107-3119.

8. Donati, C., S.C. Glotzer, and P.H. Poole, *Growing spatial correlations of particle displacements in a simulated liquid on cooling toward the glass transition.* Physical Review Letters, 1999. **82**(25): p. 5064-5067.

9. Gebremichael, Y., T.B. Schroeder, and S.C. Glotzer, *Clustering of cooperatively moving monomers in a glass-forming polymer liquid.* Abstracts of Papers of the American Chemical Society, 2000. **220**: p. 412-PHYS.

10. Glotzer, S.C. and C. Donati, *Quantifying spatially heterogeneous dynamics in computer simulations of glass-forming liquids.* Journal of Physics-Condensed Matter, 1999. **11**(10A): p. A285-A295.

11. Bennemann, C., C. Donati, J. Baschnagel, and S.C. Glotzer, *Growing range of correlated motion in a polymer melt on cooling towards the glass transition.* Nature, 1999. **399**(6733): p. 246-249.

12. Glotzer, S.C., V.N. Novikov, and T.B. Schroeder, *Time-dependent, four-point density correlation function description of dynamical heterogeneity and decoupling in supercooled liquids.* Journal of Chemical Physics, 2000. **112**(2): p. 509-512.

13. Schroeder, T.B., S. Sastry, J.C. Dyre, and S.C. Glotzer, *Crossover to potential energy landscape dominated dynamics in a model glass-forming liquid.* Journal of Chemical Physics, 2000. **112**(22): p. 9834-9840.

14. Bengtzelius, U., W. Gotze, and A. Sjolander, *Dynamics of Supercooled Liquids and the Glass-Transition.* Journal of Physics C-Solid State Physics, 1984. **17**(33): p. 5915-5934.

15. Kob, W. and H.C. Andersen, *Testing Mode-Coupling Theory for a Supercooled Binary Lennard- Jones Mixture - the Van Hove Correlation-Function.* Physical Review E, 1995. **51**(5): p. 4626-4641.

16. Sastry, S., P.G. Dcbcnedetti, and F.H. Stillinger, *Signatures of distinct dynamical regimes in the energy landscape of a glass-forming liquid.* Nature, 1998. **393**(6685): p. 554-557.

17. Bennemann, C., W. Paul, J. Baschnagel, and K. Binder, *Investigating the influence of different thermodynamic paths on the structural relaxation in a glass-forming polymer melt.* Journal of Physics-Condensed Matter, 1999. **11**(10): p. 2179-2192.

18. Bennemann, C., J. Baschnagel, and W. Paul, *Molecular-dynamics simulation of a glassy polymer melt: Incoherent scattering function.* European Physical Journal B, 1999. **10**(2): p. 323-334.

226

19. Bennemann, C., J. Baschnagel, W. Paul, and K. Binder, *Molecular-dynamics simulation of a glassy polymer melt: Rouse model and cage effect.* Computational and Theoretical Polymer Science, 1999. **9**(3-4): p. 217-226.

20. Bennemann, C., W. Paul, K. Binder, and B. Dunweg, *Molecular-dynamics simulations of the thermal glass transition in polymer melts: alpha-relaxation behavior.* Physical Review E, 1998. **57**(1): p. 843-851.

21. Deppe, D.D., R.D. Miller, and J.M. Torkelson, *Small molecule diffusion in a rubbery polymer near T-g: Effects of probe size, shape, and flexibility.* Journal of Polymer Science Part B-Polymer Physics, 1996. **34**(17): p. 2987-2997.

22. Sillescu, H., *Translation-rotation paradox for diffusion in fragile glass-forming liquids - Comments.* Physical Review E, 1996. **53**(3): p. 2992-2994.

23. Hall, D.B., A. Dhinojwala, and J.M. Torkelson, *Translation-rotation paradox for diffusion in glass-forming polymers: The role of the temperature dependence of the relaxation time distribution.* Physical Review Letters, 1997. **79**(1): p. 103-106.

24. Stillinger, F.H. and J.A. Hodgdon, *Translation-Rotation Paradox for Diffusion in Fragile Glass- Forming Liquids.* Physical Review E, 1994. **50**(3): p. 2064-2068.

25. Stillinger, F.H., *Shear viscosity and diffusion in supercooled liquids*, in *Supercooled Liquids.* 1997. p. 131-139.

26. Bohmer, R., R.V. Chamberlin, G. Diezemann, B. Geil, A. Heuer, G. Hinze, S.C. Kuebler, R. Richert, B. Schiener, H. Sillescu, H.W. Spiess, U. Tracht, and M. Wilhelm, *Nature of the non-exponential primary relaxation in structural glass-formers probed by dynamically selective experiments.* Journal of Non-Crystalline Solids, 1998. **235**: p. 1-9.

27. Wendt, H. and R. Richert, Phys. Rev. B, 2000. **61**: p. 1722.

28. Yang, M. and R. Richert, *Observation of Heterogeneity in the Nanosecond Dynamics of a Liquid.* J. Chem. Phys., 2001. **115**(6): p. 2676-2680.

29. Richert, R., J. Phys. Chem. B, 1997. **101**: p. 6323.

30. Schmidt-Rohr, K. and H.W. Spiess, Phys. Rev. Lett., 1991. **66**: p. 3020-3023.

31. Weeks, E.R., J.C. Crocker, A.C. Levitt, A. Schofield, and D.A. Weitz, *Three-dimensional direct imaging of structural relaxation near the colloidal glass transition.* Science, 2000. **287**(5453): p. 627-631.

32. Kegel, W.K. and A. van Blaaderen, *Direct observation of dynamical heterogeneities in colloidal hard-sphere suspensions.* Science, 2000. **287**(5451): p. 290-293.

33. Russina, M., E. Mezei, R. Lechner, S. Longeville, and B. Urban, *Experimental evidence for fast heterogeneous collective structural relaxation in a supercooled liquid near the glass transition.* Physical Review Letters, 2000. **84**(16): p. 3630-3633.

34. Glotzer, S., *Colloids Reinforce Glass Theory.* Physics World, 2000(April): p. 22-23.

35. Allegrini, P., J.F. Douglas, and S.C. Glotzer, *Dynamic entropy as a measure of caging and persistent particle motion in supercooled liquids.* Physical Review E, 1999. **60**(5): p. 5714-5724.

36. Gebremichael, Y., F. Starr, T. Schroeder, and S. Glotzer, *Spatially Correlated Dynamics in a Simulated Glass-forming Polymer Melt: Analysis of Clustering Phenomena.* Phys. Rev. E **64**, 051503(1–13) (2001).

37. Dasgupta, C., A.V. Indrani, S. Ramaswamy, and M.K. Phani, *Is There a Growing Correlation Length near the Glass-Transition.* Europhysics Letters, 1991. **15**(3): p. 307-312.

38. Franz, S., C. Donati, G. Parisi, and S.C. Glotzer, *On dynamical correlations in supercooled liquids.* Philosophical Magazine B-Physics of Condensed Matter Statistical Mechanics Electronic Optical and Magnetic Properties, 1999. **79**(11-12): p. 1827-1831.

39. Donati, C., S. Franz, S. Glotzer, and G. Parisi, preprint.

40. Lacevic, N., T. Schroeder, F. Starr, V. Novikov, and S. Glotzer, preprint.

41. Poole, P.H., C. Donati, and S.C. Glotzer, *Spatial correlations of particle displacements in a glass-forming liquid.* Physica A, 1998. **261**(1-2): p. 51-59.

42. Starr, F., S. Sastry, J. Douglas, and S. Glotzer, preprint.

43. Dzugutov, M., S.I. Simdyankin, and F.H.M. Zetterling, *Decoupling of diffusion from structural relaxation and spatial heterogeneity in a supercooled simple liquid.* cond-mat/0109057, 2001.

44. van Zon, R. and J. Schofield, *Mode coupling theory for multi-point and multi-time correlation functions.* cond-mat/0108029, 2001.

Chapter 17

Nanoscopic Heterogeneities in the Thermal and Dynamic Properties of Supercooled Liquids

Ralph V. Chamberlin

Department of Physics and Astronomy, Arizona State University,
Tempe, AZ 85287–1504

The theory of small system thermodynamics is applied to the Ising model *plus* kinetic energy, yielding a partition function for supercooled liquids. Several features near the glass temperature can be attributed to the resulting nanothermodynamic transition. A mean-field-like equilibrium energy exhibits Curie-Weiss-like behavior that provides an explanation for the Vogel-Tamman-Fulcher (VTF) law. Because this energy reduction is intensive, essentially independent of system size, the basic thermodynamic unit (called aggregate) subdivides into smaller regions (called clusters) lowering the net internal energy. The intensive energy reduction also yields relaxation rates that vary exponentially with *inverse* size, which when combined with the distribution of aggregate sizes provides an explanation for the Kohlrausch-Williams-Watts (KWW) law. Standard fluctuation theory gives a quantitative connection between the spectrum of response and the measured specific heat. Characteristic length scales from the model are related to direct measurements.

Perhaps the two most distinctive features in the behavior of supercooled liquids are the non-Arrhenius activation as a function of temperature T, and the nonexponential relaxation as a function of time t. Near the glass temperature T_g, measurements are usually analyzed empirically in terms of the Kohlrausch-Williams-Watts (*1*,*2*) (KWW) stretched exponential $\Phi(t) \propto \exp[-(t/\tau)^\beta]$, with the Vogel-Tamman-Fulcher (*3*,*4*) (VTF) law for the characteristic relaxation time $\tau \propto \exp[B/(T-T_0)]$. Here β is the KWW stretching exponent, Bk_B (with k_B Boltzmann's constant) gives an energy scale for activation, and T_0 is the Vogel temperature. Simple systems would have single exponential relaxation ($\beta=1$) and Arrhenius activation ($T_0=0$), but most supercooled liquids have $\beta<1$ and $T_0>0$. Although the KWW and VTF laws have been used to characterize thousands of measurements on hundreds of substances, there is still no widely accepted explanation for either formula. In fact it is still debated whether the behavior near T_g signifies a true phase transition, or merely a dynamical freezing. Indeed, the search for a fundamental theory of the glass transition has been called the "deepest and most interesting unsolved problem in solid state theory" (*5*).

Our approach (*6*) is based on the elementary observation that the VTF law from supercooled liquids is suggestively similar to the Curie-Weiss law from magnetism (*7*), which motivates us to look for a mean-field-like solution for the activation energies in supercooled liquids. This similarity seems never to have been noted before, probably because the Curie-Weiss law is usually applied to susceptibilities not activation energies, and in the mean-field approximation there is no energy reduction above the transition. However, these standard results are valid only in the limit of an infinite thermodynamic system, finite systems have a non-zero Curie-Weiss-like energy reduction above the transition, which provides a physical explanation for the VTF law.

The theory of small system thermodynamics ("nanothermodynamics") was developed by Terrell Hill about 40 years ago (*8-11*). Here we add two new features to obtain a partition function for supercooled liquids. First we show that the equilibrium energy of an isolated system is equivalent to the mean-field result, and then we use standard fluctuation theory to obtain the distribution of activation energies in bulk samples. The input parameters for the model include a net interaction energy (ε_0) between particles which form strongly interacting clusters, and a fraction of nearly-free particles (*f*) between the clusters. As a function of temperature, the Curie-Weiss-like energy yields the VTF law with $T_0 \approx \varepsilon_0/k_B$ and $B \sim T_0/f$, but finite-size effects cause deviations from the VTF law that are observed in the response of several systems (*12-15*). Similarly, the distribution of activation energies yields KWW-like behavior, but again finite-size effects cause deviations that match the measured response, including the excess wing and other features at very high frequencies (*16-20*). Furthermore, the model yields quantitative agreement with measured correlation lengths, and a connection between the response spectrum and the specific heat (*21-26*). Thus, the model provides a physical basis for both the KWW and VTF laws; while the critical behavior can be attributed to a thermal transition about a midpoint temperature T_M that is broadened by finite-size effects.

Thermodynamic heterogeneity

Most recent experimental evidence favors heterogeneity as the main source of complexity in the dynamics of condensed matter (27). Techniques which demonstrate the heterogeneous nature of the slow response include: nuclear magnetic resonance (28-30), chemical-probe bleaching (31) and solvation dynamics (32). Detailed analysis of several techniques (33,34) indicates that the intrinsic local relaxation is consistent with simple-exponential response, so that the entire spectral width can be attributed to a heterogeneous distribution of local relaxation times. From this width, dynamical correlation lengths of 1-5 nm near the glass transition have been deduced (35-37). Indeed, direct measurements (38-41) have confirmed that different relaxation times coincide with distinct spatial regions having an average radius of 1-5 nm. Dynamic heterogeneity has also been found in computer simulations (42), interacting hard spheres (43), and colloidal suspensions (44,45).

Here we focus on evidence indicating that the primary response in condensed matter involves *thermo*dynamic heterogeneity. Several specific heat measurements have shown that energy flows slowly from the thermal bath into the slow degrees of freedom. One technique involves monitoring the temperature change of the sample (ΔT) as a function of time after applying a heat pulse (46-50). At short times, ΔT rises sharply as the heat diffuses rapidly from the heater through the sample to the thermometer. At long times, ΔT returns exponentially to zero as heat flows out of the sample via a weak link to the cryostat. At intermediate times, ΔT relaxes non-exponentially as heat flows slowly from the thermal bath into the sample's slow degrees of freedom. Indeed, when ΔT is inverted to give the excess heat in the thermal bath, this enthalpy relaxation is similar to the dielectric and mechanical response in the sample. Likewise, if the amplitude and phase of ΔT is monitored as a function of heater frequency, the specific heat loss spectrum is similar to other loss spectra from the sample (51-53).

A technique that simultaneously establishes the weak thermal coupling and the heterogeneity in the primary response is nonresonant spectral hole burning, NSHB, Fig. 1. NSHB utilizes a large-amplitude pump oscillation to add energy to the slow degrees of freedom, then a probe step to measure the modification. Observation of NSHB requires that the excess energy from the pump oscillation must flow slowly out of the selected slow degrees of freedom, consistent with the inverse process of dynamic specific heat where energy added to the thermal bath flows slowly into the slow degrees of freedom. The intrinsic behavior of the response is deduced from the measured modification. For homogeneous response [Fig. 1(a)] the entire spectrum is shifted uniformly toward shorter times, whereas for heterogeneous response [Fig. 1(b)] a spectral hole develops. Because NSHB uses the responding degrees of freedom as their own probe, it is the most direct technique for investigating the nature of the net response from a bulk sample, but it cannot determine the length scale of the heterogeneity. The inset of Fig. 1 (b) showing a cartoon sketch of

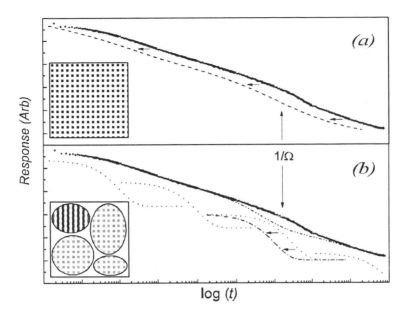

Figure 1. NSHB characteristics for (a) homogeneous and (b) heterogeneous response after a pump oscillation at frequency $\Omega/2\pi$.

spatially-localized spectrally-distinct degrees of freedom is established by other types of measurements that utilize local-probes.

NSHB was first applied to the nonexponential dielectric relaxation of supercooled liquids (*54*), Fig. 2. A simple model (*55*, inset of Fig. 2) consists of a "box" for each set of slow degrees of freedom, with a local temperature T_i, response rate $1/\tau_i$, and corresponding recovery rate $1/(\tau_i+\gamma)$ via a set of thermal resistors coupled to the thermal bath. Here $1/\gamma$ is a uniform limit to how fast heat can flow out of the slow degrees of freedom, which can be attributed to a thermal bottleneck. Deduced values of $1/\gamma$ are within experimental error of the average alpha response rate, consistent with the nearly uniform rate seen for aging of the alpha wing after a thermal quench (*56*). The model provides good agreement with all prominent features in the NSHB behavior (solid curves in Fig. 2). Dielectric NSHB has also been observed in a relaxor ferroelectric (*57*) and an ionic glass (*58*); while magnetic NSHB has been measured in a spin glass (*59*) and two crystalline ferromagnets (*60*). In every case, the observation of spectral holes indicates that the nonexponential primary response involves thermodynamic heterogeneity. This picture of spectrally-distinct regions that are weakly coupled to a thermal bath is the basis of the nanothermodynamic model presented here.

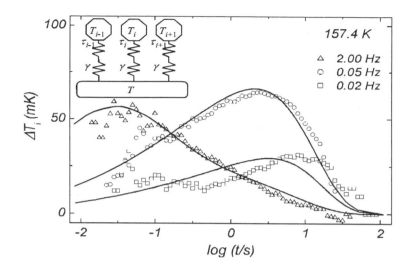

Figure 2. Spectral holes in supercooled propylene carbonate after pump oscillations at three different frequencies (*54*). The bath temperature was held constant at 157.4 K. Changes in local relaxation time are plotted as changes in local temperature using the measured activation energy of the alpha response peak. The inset shows the "box" model (*55*) used to calculate the solid curves.

Nanothermodynamic Model

We characterize the local interactions between slow degrees of freedom using the Ising model. In the Ising model, the ith degree of freedom may be represented by a "particle" with a state variable (σ_i) that is either "up" (+1), or "down" (-1). For magnetic systems, σ_i represents the up or down orientation of each Ising spin. Other systems with two degrees of freedom, such as a lattice gas or binary alloy, map directly onto the Ising model. For supercooled liquids, the binary degrees of freedom may correspond to the two lowest energy levels in the interaction between molecules. Often the energy levels depend on the electron alignments in the molecules, as in the case of bonding and anti-bonding states, which would also map to the Ising model. Otherwise the Ising model is just the first step towards characterizing the actual interaction between complex molecules, for which we assume that each molecule with multiple degrees of freedom may be represented by multiple Ising-like "particles."

Consider a cluster containing a total of m particles, with a uniform interaction energy of $\varepsilon_0/2z$ between each particle and its z nearest neighbors in the cluster. The net interaction energy per particle is $V_m = -(\varepsilon_0/2zm)\sum_{<ij>}\sigma_i\sigma_j$, where $<ij>$ denotes the sum over all pairs of interacting particles. In terms of the number of up particles (ℓ)

and the number of pairs of interacting up particles (ℓ_{++}), the energy may be written as (61) $V_m(\ell,\ell_{++})=-(\varepsilon_0/4)(8\ell_{++}/zm - 4\ell/m+1)$. Most of the difficulty in evaluating $V_m(\ell,\ell_{++})$ comes from the pair-interaction term ℓ_{++}. In the mean-field approximation, this difficulty is avoided by using the average value $\overline{\ell_{++}}/m \approx (z/2)(\ell/m)^2$, leaving $\overline{V_m}(\ell) \approx -(\varepsilon_0/4)(2\ell/m - 1)^2$.

Equilibrium energy of finite clusters

Although the mean-field approximation is usually associated with long-ranged interactions, in fact an identical expression is obtained for nearest-neighbor interactions in an isolated cluster (microcanonical ensemble) with two basic assumptions. 1) Net momentum is conserved; thus changes in ℓ occur only through contact to neighboring clusters, whereas ℓ-conserving changes in the local configuration (ℓ_{++}) occur internally. 2) Net energy per particle $[\varepsilon_m(\ell)]$ is conserved by a balance between potential and kinetic energies; thus configurations with low potential energy have high kinetic energy and vice versa, so that the virial theorem for any realistic potential near a stable equilibrium yields $\varepsilon_m(\ell)=2\overline{V_m}(\ell)$. In the Ising model, kinetic-energy-carrying "demons" have been used to maintain the microcanonical ensemble $(62,63)$. Using the fundamental postulates of statistical mechanics - that all allowed states are equally likely and that the equilibrium of a finite system is found from its time-averaged properties - the net energy per particle is $\varepsilon_m(\ell)\approx-(\varepsilon_0/2)[4\ell(\ell-1)/m(m-1) - 4\ell/m + 1]$.

One way to demonstrate the validity of $\varepsilon_m(\ell)$ is to examine the interaction energies of some specific examples. Consider four Ising particles $(m=4)$ arranged in a square ring so that each particle interacts with only its two nearest neighbors. For two up particles $(\ell=2)$, one set of four-fold degenerate configurations has both aligned particles on the same side of the square $(\ell_{++}=1)$ giving $V_4(2,1)=0$; and a doubly-degenerate set has the aligned particles at diagonal corners $(\ell_{++}=0)$ giving $V_4(2,0) = +\varepsilon_0/4$. Because all six configurations are equally likely, the equilibrium potential energy per particle is $\overline{V_4}(2)=[4(0)+2(\varepsilon_0/4)]/6 = \varepsilon_0/12$. Thus, even for a small cluster with only two nearest neighbors the net energy is in full agreement with $\varepsilon_4(2)=\varepsilon_0/6$. Similarly, for a cluster containing eight particles, Fig. 3 shows that the long-ranged and time-averaged interaction energies are identical; and for large clusters, when $(\ell-1)/(m-1)\approx\ell/m$, this energy approaches the standard mean-field result. In fact, if the averaging process conserves the average alignment ($\overline{\ell}$ =const) instead of the specific state (ℓ=const), then the finite-size mean-field result is always accurate. Indeed, this $\varepsilon_m(\ell)=-(\varepsilon_0/2)(2\ell/m-1)^2$ gives good agreement with the internal energies found from measurements on many materials. Thus, when describing slow response or equilibrium properties, the local degrees of freedom which render the 3-D Ising model mathematically intractable (64) are integrated out via thermal averaging. Adding kinetic energy to the Ising model also facilitates normal-mode excitations and the microcanonical ensemble, making the model more realistic.

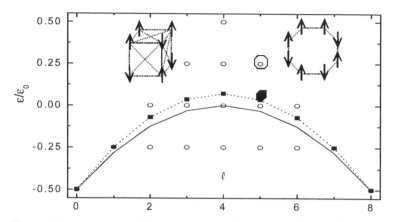

Figure 3. Energy per particle as a function of alignment ℓ in the Ising model for: (o) a ring with two neighbors per site, and (■) a cube with interactions between all particles. The solid curve is from the finite-size mean-field result. The dashed curve is from the average of the ring energies, or equivalently from the Ising model plus kinetic energy at fixed ℓ in the microcanonical ensemble.

Unrestricted thermal fluctuations

Now let the cluster interact weakly with its environment. The exchange of momentum allows all possible values of ℓ and the exchange of energy defines the temperature T. Using the binomial coefficient for the degeneracy of each state, the partition function for clusters of size m is

$$\Delta_m = \sum_{\ell=0}^{m} \frac{m!}{(m-\ell)!\ell!} \exp[-m\varepsilon_m(\ell)/k_B T] \tag{1}$$

For analytic evaluation it is convenient to make a change of variables to $L=(2\ell/m-1)$, leaving $\varepsilon(L)=-\frac{1}{2}\varepsilon_0 L^2$. Note that the magnitude ($|L|$) is the usual Ising order parameter, and that the energy is a maximum when there is no order $\varepsilon(0)=0$, and a minimum when all particles are either up or down $\varepsilon(\pm 1) = -\varepsilon_0/2$. For sufficiently large m, using Stirling's approximation for the factorials and converting the sum to an integral, the partition function becomes $\Delta_m \approx \int_{-1}^{+1} dL \exp[-mf(L)/k_B T]$ where the free energy per particle is $f(L)=\varepsilon(L) + \frac{1}{2}k_B T\{(1+L)\ln[\frac{1}{2}(1+L)]+(1-L)\ln[\frac{1}{2}(1-L)]\}$. For an infinite cluster ($m\rightarrow\infty$), only the minimum free energy contributes to the partition function, and setting $f'(L)\equiv\partial f(L)/\partial L=0$ yields the usual transcendental equation for the order parameter, $L_\infty=\tanh[L_\infty(\varepsilon_0/k_B T)]$. This solution exhibits the well-known Curie-Weiss transition at $T=\varepsilon_0/k_B$, with $L_\infty=0$ and hence $\varepsilon(L_\infty)=0$ for $T>\varepsilon_0/k_B$. However, this standard result is valid only for infinite systems. We now show that a finite system can lower its energy by subdividing into small clusters.

Continuing with the integral representation of the partition function, the average energy for a cluster of size m (from $k_B T^2 \partial\ln \Delta_m / \partial T$) is

$$\overline{E_m} \approx \int_{-1}^{+1} dL(-\tfrac{1}{2}m\varepsilon_0 L^2)\exp[-mf(L)/k_B T]/\Delta_m \tag{2}$$

For large but finite m, the integrals may be approximated by a steepest-descents procedure, where the argument in the exponent of each integrand is expanded about its maximum. Specifically for the partition function in the denominator, at $T > \varepsilon_0/k_B$ where $L_\infty = 0$, $f(L) \approx f(0) + f''(0)\tfrac{1}{2}L^2$ gives $\Delta_m \approx 2^m\sqrt{2\pi/m(1-\varepsilon_0/k_B T)}$. However the extra factor of L^2 in the numerator yields a different transcendental equation $L_m \approx \tanh[L_m(\varepsilon_0/k_B T) + 2/(mL_m)]$, with a non-zero order parameter $|L_m| \approx \sqrt{2/m(1-\varepsilon_0/k_B T)}$ and non-zero energy reduction $\overline{E_m} \approx -\tfrac{1}{2}\varepsilon_0/(1-\varepsilon_0/k_B T)$ *above* the transition. Indeed, the energy reduction and order parameter are non-zero at all temperatures [Fig. (4)]; and the integral approximations become *increasingly* accurate with increasing m. However, note that $|L_m| \propto m^{-1/2}$ and that the total energy is intensive, thereby returning the standard mean-field result that an infinite cluster at $T > \varepsilon_0/k_B$ is unordered, with negligible energy reduction. Although the total energy reduction of a typical cluster is small, comparable to a single pair of aligned particles, the very large number of nanoscopic clusters yields a macroscopic energy reduction for the whole sample.

Another way to picture the physical mechanism driving cluster formation is as follows. An infinite cluster is unordered at $T > \varepsilon_0/k_B$, so its net energy per particle is $\varepsilon_m(m/2) = 0$. Thus, a bulk sample can lower its net energy by subdividing into small,

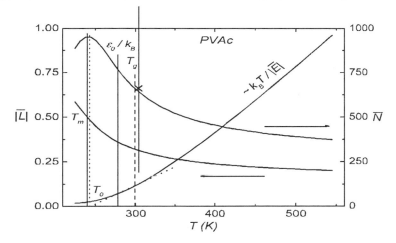

Figure 4. Average local order parameter, aggregate size, and inverse activation energy for polyvinyl acetate as deduced from a fit to the temperature-dependent dielectric loss peak. Various temperatures of interest are shown. The X marks the dynamic correlation size (with its standard deviation bar) as determined by a simple average of five different measurements from the literature (*36, 38-40*).

more highly-ordered clusters, separated by surfaces where the time-averaged interaction is temporarily zero, $\overline{\sigma_i \sigma_j} = 0$. In fact, the same type of dynamical averaging that yields the mean-field energy within a cluster when coherent, yields zero interaction between clusters when incoherent. Presumably due to interfacial tension, or some other details in the averaging process that are not contained in the basic model presented here, the clusters are compact and spheroidal. The ability of interactions to adjust themselves to lower the net energy of the sample is a common feature of fast degrees of freedom in the adiabatic approximation.

Now let each cluster also exchange particles with its environment, allowing fluctuations in cluster size m. Returning to the explicit summation for Δ_m in Eq. (1), the partition function for clusters with variable size is

$$\Gamma = \sum_{m=m_0}^{\infty} \Delta_m e^{m\mu/k_B T}, \tag{3}$$

where μ is the chemical potential. Here the summation starts at a minimum cluster size m_0, which is adjusted to the integer value that gives best agreement with measured response. Most of the values in Table I are slightly less than the $m_0 \approx 13$ for a central site and its nearest-neighbor shell of randomly packed spheres, as might be expected for real molecules with steric constraints. The m_0 for salol is about twice the other values, possibly due to dimerization, or a second type of interaction. Although Eq. (3) resembles the grand canonical ensemble, in fact Γ contains only the intensive variables μ and T [pressure (p) can be ignored for liquids that are effectively incompressible], thus allowing the clusters to fluctuate in both energy *and* size. If the summation were extended to $m_0=1$, then cluster sizes

Table I. **Fit parameters and deduced quantities from best fits to the temperature-dependent dielectric-loss peak of: salol, propylene carbonate (PC), 2-methyltetra- hydrofuran (MTHF), polyvinyl acetate (PVAc), glycerol, and propylene glycol (PG). The average cluster size $\overline{m}$ and number of clusters per aggregate $\overline{n}$ have a weak temperature dependence; values given are at the transition mid point, T_M.**

substance	ε_0/k_B (K)	f	f_0 (THz)	m_0	$\overline{m}$	$\overline{n}$	$\dfrac{\Delta c_p}{\overline{C_p/N}}$
salol	257	0.079	0.014	24	58	2.6	3
PC	182	0.086	0.021	12	38	5.1	3
MTHF	100	0.086	0.031	9	34	6.5	2
PVAc	272	0.050	0.046	9	48	19	1/2
glycerol	121	≈ 0	28	9	75	220	3
PG	105	≈ 0	7.7	8	80	240	3

would fluctuate without restriction. Such completely open ("generalized") ensembles are rarely applied to macroscopic systems where at least one extensive variable must be used to fix the size of the system. However, for bulk samples with unrestricted internal correlations, all extensive variables must be free to fluctuate; and a key feature in the theory of nanothermodynamics is that since small systems have large fluctuations only the appropriate ensemble yields correct behavior.

The ensembles that are relevant to our model can be clarified by a brief review of nanothermodynamics. Consider a small system ("box") surrounded by its environment, Fig. 5. In the microcanonical ensemble, the box is completely isolated from its surroundings, so that its environmental variables are particle number N, volume v, and energy E. Quantitative thermostatistics is based on this ensemble, and the assumption that all states which conserve N, v, and E are equally likely. In the canonical ensemble, the box exchanges heat with its surroundings, so that energy fluctuates and T replaces E as one of the environmental variables. In the grand canonical ensemble, the box also exchanges particles with its surroundings, so that N fluctuates leaving μ, v, and T as the environmental variables. The final step is to also allow flexibility in the volume of the box, leaving a completely open "generalized" ensemble where all extensive variables are free to fluctuate. Most textbooks state that the it makes no difference which ensemble is used, but this is true only for infinite systems where fluctuations can be ignored. In fact, because it is difficult to imagine how a local fluctuation in a bulk sample could exchange energy freely with its environment without changing its size (N and/or v), the

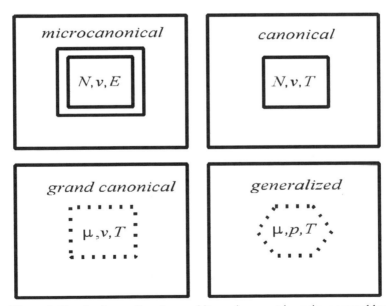

Figure 5. Sketch of a small system and its environment in various ensembles.

canonical and grand canonical ensembles are generally not valid for any response that involves *thermodynamic* heterogeneity. Thus only two basic ensembles are relevant to the internal dynamics of bulk materials: microcanonical for the fast degrees of freedom that change without coupling to their surroundings, and generalized for the slow degrees of freedom that are coupled to their environment.

The partition function of Eq. (3) is incomplete because it assumes that all particles are in clusters, and that all clusters exchange thermal properties equally across the entire sample. We compensate for these errors by considering aggregates comprised of a finite number of clusters n, and a fraction f of nearly-free particles between the clusters. Because the clusters fluctuate without restriction, no cluster can be identified by its size, shape, or location. Thus, on time scales long compared to the fluctuations, all clusters in an aggregate are statistically indistinguishable (8). The partition function for aggregates with all possible number of clusters is

$$\Upsilon = \sum_{n=1}^{\infty} \frac{\Gamma^n}{n!} \exp(nf\,\overline{m}\mu/k_B T) \tag{4}$$

where $\overline{m} = k_B T \partial \ln \Gamma / \partial \mu$ is the average cluster size, and $n(1+f)\,\overline{m}$ is the total number of particles in an aggregate containing n average-sized clusters. The partition function now includes a number of nearly-free particles ($nf\,\overline{m}$), and an interaction between nearby clusters in the form of the direct exchange of thermal properties. Note that the fraction of nearly-free particles is an effectively constant geometric ratio, and that the usual filling factor is equal to $1/(1+f)$. For random close-packed uniform hard spheres $f = 0.57$, whereas for realistic clusters which vary in size and need not be spherical $f < 0.57$, see Table I.

Thermal equilibrium is found from Gibbs' variational principle (61) by adjusting the net chemical potential to maximize Υ with all other variables fixed, $(\partial \Upsilon / \partial \mu)_{f,T} = 0$. This is a new type of "generalized" ensemble with f and T as the thermodynamic parameters, so that all the extensive parameters and also μ can fluctuate, which allows the coexistence of solid-like clusters and nearly-free particles. Thus, as a function of reduced temperature $(k_B T/\varepsilon_0)$, f alone governs both the average number of particles in a cluster ($\overline{m}$) *and* the average number of clusters in an aggregate ($\overline{n} = \Gamma \partial \ln \Upsilon / \partial \Gamma$). Glycerol and propylene glycol differ in that they have a vanishingly small fraction of free particles, possibly due to the hydrogen bonding between molecules. For these "strong" liquids we set $f = 0$ in Eq. (4) and adjust $\mu/k_B T$ in Eq. (3) to the constant value which best fits the data, yielding a more traditional type of μ-T generalized ensemble.

Some features in the dynamics of our model can be illustrated by comparison to molecular-dynamics simulations. To avoid crystallization, most simulations utilize a mixture of two types of molecules. We also consider two types of molecules; but because $\sigma_i = \pm 1$ is due to *intra*molecular degrees of freedom, there is *dynamical* heterogeneity in the interaction strength, even between otherwise identical molecules at fixed separation. In addition to being more realistic for single-component molecular liquids, the values of σ_i are free to adjust themselves to minimize the net free energy of the sample, including the formation of cluster surfaces where $\overline{\sigma_i \sigma_j}$ =0. Several simulations (*42-45*) have found that excess motion tends to concentrate into a small number of mobile particles, while the remaining particles are more static. In fact, a particular computer simulation (*65*) of a Lennard-Jones liquid yields a relatively constant fraction of mobile particles (0.055±0.005), similar to the fractions listed in Table I. However, we emphasize that in our model the nearly-free particles are of secondary importance to the primary response which involves the cooperative activation of all the particles in each aggregate. Furthermore, because of the rapid exchange with clusters, any direct signature of the nearly-free particles in the slow dynamics may not be obvious.

Cluster and aggregate response

The hierarchy of response in a sample is governed by the interaction energies and thermalization times of various degrees of freedom. In our model, the degrees of freedom are parameterized by the local alignment (ℓ_{++}), net alignment (ℓ), cluster size (m), and number of clusters in an aggregate (n). The two main features in the response of supercooled liquids are the high-frequency behavior (boson peak and beta wing), and the broad spectrum of primary (alpha) response. We ascribe these features to the local interaction energies within clusters, and the thermal activation of aggregates, respectively.

Changes in ℓ_{++} that conserve the net alignment, energy, and size of a cluster need no external coupling, hence they occur rapidly with the *cluster* in the microcanonical ensemble. These changes involve the local interaction energy, which using Planck's constant (h) gives frequencies of $\varepsilon_0/h \sim 10^{12}$ Hz. Assuming a damped harmonic oscillator at the frequency corresponding to the interaction energy of each local configuration, with no adjustment for the spectral positions or relative weights of the configurations, the resulting set of broadened resonances match the behavior of the boson peak and beta wing from 10^9-10^{13} Hz (*66*). It is this fast adiabatic averaging of ℓ_{++} that yields the mean-field equilibrium energy within each cluster. At longer times, to ensure unrestricted exchange of thermal properties between clusters, all *clusters* in each aggregate form a generalized ensemble.

Changes in ℓ, m, and n that conserve the net alignment, energy, and size of an aggregate need no external coupling, hence they proceed with the *aggregate* in the microcanonical ensemble. These changes occur on intermediate time scales, but because the net properties of the aggregate are conserved, their influence on the bulk behavior is relatively weak, possibly appearing as some subtle features near the minimum between the boson and alpha peaks. Because the aggregates contain typically more than a hundred degrees of freedom (particles) which are effectively uncoupled from their outside environment, each aggregate forms a distinct thermal system, yielding the experimentally observed *thermo*dynamic heterogeneity. Moreover, the empirical observation that the net internal energy of each aggregate is conserved prior to its alpha response confirms their microcanonical isolation, and justifies the use of a local "fictive" temperature for each aggregate. Finally, during the alpha response, to ensure unrestricted fluctuations as the aggregates couple to each other via thermal activation, all *aggregates* in the sample form a generalized ensemble. Because of rapid internal averaging, the net relaxation rate of each aggregate depends mainly on its equilibrium energy $\overline{E}$ and total size $N=n(1+f)m$.

Let w_N be the relaxation rate for an aggregate containing N particles. According to standard thermal fluctuation theory (67), the probability that this aggregate has an actual energy of E is $P_E \propto \exp[-\frac{1}{2}F"(\overline{E})(E-\overline{E})^2]$, where $F"(\overline{E})=(\partial^2 F/\partial E^2)_{\overline{E}}$ is the free energy curvature about its equilibrium value $F(\overline{E})$. In the usual thermodynamic limit the internal energy is extensive so that $F"(\overline{E}) \propto 1/N$; but for finite systems in the model presented here E is intensive so that $F"(\overline{E}) \sim N$. From a Maxwell relation, this can be written as $F"(E) \sim N/(T\overline{C_p})$, where $\overline{C_p}$ is the heat capacity of an average-sized aggregate. Integrating the Boltzmann-weighted activation energies over all possible fluctuations ($\int_{-\infty}^{\infty} P_E e^{E/k_B T} dE$), using a constant attempt frequency f_0, the size-dependent relaxation rate becomes (59,60)

$$w_N = 2\pi f_0 \, e^{-|\overline{E}|/k_B T} \exp(\frac{\overline{C_p}/k_B}{2N}). \tag{5}$$

Note that only the average energy appears in the first exponential (Arrhenius term), whereas size appears explicitly only in the inverse behavior in the "inverse Arrhenius" term. To a first approximation, these two inverses cancel one another to yield the usual result that small aggregates relax faster than large aggregates; but because $\overline{E}$ is not extensive, the mechanism is not the usual size-dependent activation energy. Instead, small aggregates have large thermal fluctuations which are activated by a broad spectrum of excitations, whereas large aggregates require more precisely matched activation energies. Qualitatively, the inverse size dependence converts a relatively symmetric size distribution [which is approximately a truncated Gaussian (68)] into the observed asymmetric spectrum of response, while the "inverse Arrhenius" factor shifts this asymmetric excess wing to the high-frequency side of the peak. Thus, standard thermal fluctuation theory applied to an equilibrium distribution of finite clusters yields asymmetric KWW-like behavior, without having to use an ad hoc distribution of relaxation times or an exchange process (69) to match the measured asymmetry.

Discussion

Comparison with experiments

We use the partition function Eq. (4), with Eqs. (1) and (3), to characterize the thermal and dynamic properties of supercooled liquids. Recall that the energy of an average cluster exhibits Curie-Weiss-like behavior, so that the energy of an average aggregate may be written $\overline{E} = n\overline{E}_m \approx \overline{n}\ [-\frac{1}{2}\varepsilon_0 T/(T-\varepsilon_0/k_B)]$. Using this approximate form for the activation energy in the Arrhenius law [temperature-dependent term in Eq. (5)], yields a characteristic relaxation time $\{\tau \propto \exp[B/(T-T_0)]\}$ that obeys the VTF law with $T_0 \approx \varepsilon_0/k_B$ and $B \approx \frac{1}{2}\ \overline{n}\ T_0$. However, the solid curves in Fig. 6 show that using $\overline{E} = -k_B\ \partial\ln\Upsilon/\partial(1/T)$ without mathematical approximation gives better agreement with measured response, including typical deviations from the VTF law.

The equilibrium distribution of aggregate sizes (P_N) can be obtained from the distribution of cluster sizes [$P_m = \Delta_m\exp(m\mu/k_BT)/\Gamma$] and the number of clusters in each aggregate [$P_n = \Gamma^n\exp(nf\overline{m}\ \mu/k_BT)/n!\Upsilon$]. For aggregates containing a single cluster, $n=1$, the distribution is simply $P_N = P_1P_m$. For $n>1$, the distribution contains multiple terms that are difficult to calculate exactly, e.g. when n=2, $P_N = P_2(P_mP_m + P_{m+1}P_{m-1} + \bullet\ \bullet\ \bullet)$. However, since P_m is a relatively slowly varying function of m, for simplicity we factor out one P_m and extend the remaining summation(s) over all cluster sizes, leaving $P_N \approx P_nP_m$. Figure 4 shows that the average number of particles in an aggregate deduced from P_N is consistent with measured values of the correlation length in PVAc.

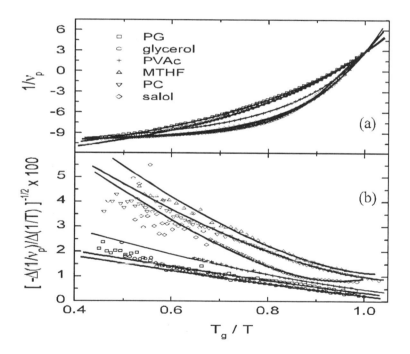

Figure 6. (a) Angell plot of the dielectric loss peaks from six substances (*12-15*). The curvature indicates non-Arrhenius behavior. (b) Stickel plot from the derivatives of the same data. The curvature indicates non-VTF behavior. The solid curves are from a best fit to the data using $\overline{E}$ as the activation energy.

The spectrum of response is obtained by combining the linearly-weighted distribution of aggregate sizes NP_N, with the size-dependent relaxation rates w_N. The net relaxation of a bulk sample is

$$\Phi(t) = \Phi_0 \sum_{N=1}^{\infty} NP_N e^{-tw_N} \qquad (6)$$

where Φ_0 is a constant amplitude prefactor. Complex response as a function of frequency (f) is given by the Fourier transform of $-d\Phi(t)/dt$,

$$\chi'(f) - i\chi''(f) = \Phi_0 \sum_{N=1}^{\infty} NP_N \frac{1}{1 + i2\pi f / w_N}. \qquad (7)$$

Figure 7 shows the best fits to several dielectric loss spectra using Eq. (7) (dashed curves), with $\overline{C_p}$ as the only adjustable parameter governing the width and shape of the response. The agreement extends well into the high frequency wing where all simple empirical formulas such as the KWW law fail. Furthermore, various features at high frequencies can be accounted for by discrete summations in the model. Specifically, the bump at ~10^6 Hz, identified as a type A beta response peak (20,70), can be attributed to the alpha response from aggregates containing two clusters with all possible sizes; while the boson peak and beta wing can be attributed (solid curves) to the set of damped harmonic resonances from the local interaction energies. Thus the alpha response comes from aggregate energies using Boltzmann's constant to obtain the activation rates, while the boson peak comes

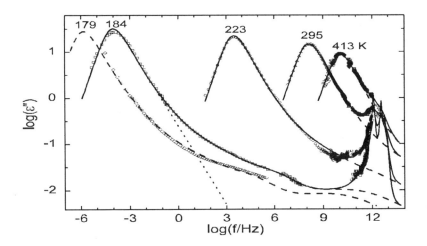

Figure 7. Dielectric loss of glycerol at five temperatures (20). Dashed curves are for aggregate activation, Eq. (7). Solid curves include a set of damped harmonic resonances corresponding to the energy of each local configuration. The straight dotted line shows the constant-slope high-frequency behavior of the KWW law.

from local interaction energies using Planck's constant to obtain the resonance frequencies, needing no adjustable parameters for the position of either peak.

Figure 8 shows a comparison between the average heat capacity per particle deduced from dielectric spectra, $\overline{C_p}/\overline{N}$, and the measured excess specific heat per molecule, Δc_p. Good quantitative agreement is obtained by multiplying $\overline{C_p}/\overline{N}$ by a rational constant (Table I) which gives the number of "particles" (binary degrees of freedom) per molecule. Most of the non-polymeric substances have 3 particles per molecule, while PVAc requires 6 monomers to achieve the same 3 degrees of freedom.

Comparison with other models

Use of the generalized ensemble for the localized degrees of freedom inside bulk materials is the main feature that distinguishes our approach from all previous models. Other novel features include using mean-field-like behavior within finite clusters, and only weak exchange of thermal properties between clusters; whereas the classic Mayer expansion (8,61) involves enumerating the discrete nearest-neighbor interactions within clusters, then using a mean-field interaction between clusters. Equations (3) and (4) are somewhat similar to the respective partition functions for lattice gas clusters and spherical aggregates on pgs. 83-90 in part II

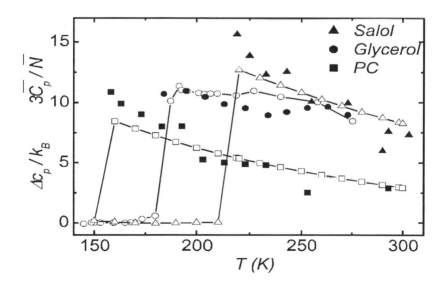

Figure 8. Measured excess specific heat (24,25) (open symbols), and average aggregate heat capacity per particle deduced from the spectra of three substances.

of Ref. (*9*). The key distinctions are: instead of a non-interacting lattice gas we use the Ising model *plus* kinetic energy to obtain an exact equilibrium energy between binary degrees of freedom in a cluster; we have aggregates of clusters instead of aggregates of individual molecules; and we assume that the clusters are dynamic enough to be indistinguishable. Furthermore, we neglect all surface effects. We have tried modified versions of Eqs. (3) and (4) which include interface terms proportional to $m^{2/3}$ and $n^{2/3}$, but these additional parameters do not significantly improve the agreement with experiments. Apparently the average clusters are large enough, and the interaction between interior particles is sufficiently similar to the interfacial interactions, that the surface free energy can be incorporated into the cluster-size cutoff m_0 and the net chemical potential μ.

The average cluster size $\overline{m}$ and average aggregate size $\overline{N}$ define *two* distinct size scales in supercooled liquids. Empirically, the two size scales are necessary for the two distinct energy scales ($k_B T_0$ and $B k_B$) in the VTF law. Several previous models for a size scale in supercooled liquids are based on the Adam-Gibbs (*71*) concept of a cooperatively rearranging region (CRR). A CRR is somewhat similar to our cluster, but there are some key distinctions. To obtain the VTF law, Adam-Gibbs assumed a linear temperature dependence for a configurational entropy in the denominator of the activation energy; whereas we obtain a mean-field-like expression for finite systems and make a connection to the Curie-Weiss law. For the slow relaxation near T_g: a relatively high entropy barrier is required for the ~5 molecules (*72,73*) in a CRR, implying ~ 10^4 discrete orientations for each molecule; whereas we use the Ising model for only two orientations per particle, so that a typical aggregate with 10^2-10^4 particles has a much lower activation energy density. Furthermore, from Fig. 4 and Table I, PVAc near T_g has $\overline{m} \approx 50$ and $\overline{N} \approx 650$ particles, corresponding to about 100 and 1300 monomers, respectively. Assuming spherical objects, this yields an average cluster radius of 1.4 nm and aggregate radius of 3.4 nm, in reasonable agreement with values deduced from dielectric fluctuations [5±2 nm (*38*) and 1.5 nm (*40*)] and spin diffusion [1.5±0.5 nm, (*39*)].

The basic Adam-Gibbs theory generally yields CRR sizes that are smaller than measured experimentally. Thus, Donth (*35*) and others (*36,37,74,75*) have considered thermal fluctuations within the cooperative regions that, like our model, vary *inversely* proportional to size and yield reasonable values for the correlation range. Unlike our model, however, they usually assume a uniform size for all correlated volumes at a given temperature, so that energy fluctuations alone are responsible for the broadened transition and spectral width. But such simple fluctuations on otherwise homogeneous volumes cannot account for the observed spectral *shape*. Specifically the high-frequency wing, which at T_g extends about 10^{12} Hz above the peak response, would require asymmetrical fluctuations of about 100 K. In contrast, we assume a distribution of aggregate sizes, whose size-dependent energy fluctuations yield good agreement with observed spectral widths *and* shapes.

The free-volume model is another approach for obtaining VTF-like behavior. The theory of Cohen and Grest (CG, *76*) is especially detailed. Like CG, we assume partially-ordered clusters with a fraction of nearly-free particles, but there are several differences. Instead of a temperature-dependent percolation of solid clusters, we assume a constant fraction of nearly-free particles so that the transition occurs within the clusters. The percolation transition in CG exhibits 1st-order-like behavior which must be broadened by heterogeneity; whereas we consider a 2nd-order-like transition which is suppressed by finite-size effects. For the VTF law, instead of a linear temperature dependence in the denominator of the activation energy [which has been questioned as the free volume nears zero (*77*)], we consider a mean-field solution which makes a connection to the Curie-Weiss law. For the KWW law CG have found $\beta > 2/3$, whereas many systems have $\beta < 2/3$. Furthermore, Figs. 6 and 7 show good agreement with typical deviations from both the VTF and KWW laws.

Much recent interest in supercooled liquids has come from the mode-coupling theory (*78*) for a dynamical transition at a T_c which is typically 20-50 K above T_g. At least for salol (*79*) and PC (*80*), T_c is within experimental error equal to the local interaction energy at $\varepsilon_0/k_B > T_g$, see Table II. However, other systems have $\varepsilon_0/k_B < T_g$, consistent with the lack of any clear critical behavior above T_g in glycerol (*81*) and PVAc (*82*). Thus, like standard Curie-Weiss behavior, various high temperature properties may extrapolate toward ε_0/k_B, while the true transition is centered about $T_M < \varepsilon_0/k_B$. Nevertheless, some dynamical precursors to the transition could occur at ε_0/k_B. In addition, some sets of data show features in the response near the minimum between the alpha and boson peaks that are not fully accounted for by our model, possibly indicating the presence of cage effects. Still, for the peak primary response, our model for a mean-field interaction within clusters remains valid as long as the sample exhibits VTF-like behavior, typically to temperatures of about twice T_g.

We have shown that finite-size effects provide a physical basis for both the VTF and KWW laws in supercooled liquids, including the most commonly observed deviations from these empirical formulas. And since VTF- and KWW-like behavior are found in a wide variety of liquids, glasses, polymers and crystals (*60*), we speculate that nanothermodynamics may also be relevant to these systems.

Acknowledgments

I thank S. Benkhof, F. Kremer, P. Lunkenheimer, S. R. Nagel, M. Oguni, R. Richert and B. Scheiner for providing me with the original data presented here. I also thank Roland Böhmer, Burkhard Geil, Jeppe Dyre, Terrell Hill, Ranko Richert, Kevin Schmidt, and George Wolf for enlightening discussions. Special acknowledgment is given to Prof. Hans Sillescu in honor of his 65th birthday. This research was funded by the NSF, grant number DMR-9701740.

Table II. Glass (T_g), Vogel (T_0), mode-coupling (T_c), transition mid-point (T_M) and local interaction energy ε_0/k_B temperatures for the substances identified in Table I.

substance	T_g (K)	T_0 (K)	T_c (K)	T_M (K)	ε_0/k_B (K)
salol	217	175-224	256-268	225	257
PC	163	132-153	187	160	182
MTHF	91.5	69.6	?	89.3	100
PVAc	299	243-250	288-306	240	272
glycerol	188	135	225-305	107	121
PG	167	127	205-251	92.5	105

References

1. Kohlrausch, R. *Pogg. Ann. Phys.* **1854**, *91*, 56.
2. Williams, G.; Watts, D. C., *Trans. Faraday Soc.* **1970**, *66*, 80.
3. Vogel, H. *Phys. Z.* **1921**, *22*, 645.
4. Fulcher, G. S. *J. Am. Ceram. Soc.* **1925**, *8*, 339.
5. Anderson P. W. *Science* **1995**, *267*, 1615.
6. Chamberlin, R. V. *Phys. Rev. Lett.* **1999**, *82*, 2520; and *Nature* **2000**, *408*, 337.
7. Weiss, P. *Comptes Rendus* **1906**, *143*, 1136.
8. Hill, T. L. *Statistical Mechanics*; Dover: New York, **1987**; and *An Introduction to Statistical Thermodynamics*; Dover: New York, **1986**.
9. Hill, T. L. *J. Chem. Phys.* **1962**, *36*, 3182; and *Thermodynamics of Small Systems, Parts I and II*; Dover: New York, **1994**.
10. Hill, T. L.; Chamberlin, R. V. *Proc. Natl. Acad. Sci. USA* **1998**, *95*, 12779.
11. Hill, T. L. *Nano Lett.* **2001**, *1*, 111; *ibid* **2001**, *1*, 273.
12. Stickel, F.; Fischer, E. W.; Schönhals, A.; Kremer, F. *Phys. Rev. Lett.* **1994**, *73*, 2936.
13. Stickel, F.; Fischer, E. W.; Richert, R. *J. Chem. Phys.* **1996**, *104*, 2043.
14. Stickel, F.; Thesis, Mainz University; Shaker: Aachen, Germany **1995**.
15. Richert, R.; Wagner, H. *Sol. St. Ionics* **1998**, *105*, 167.
16. Dixon, P. K.; Wu, L.; Nagel, S. R.; Williams, B. D.; Carinin, J. P. *Phys. Rev. Lett.* **1990**, *65*, 1108.
17. Kudlick, A.; Benkhof, S.; Lenk, R.; Rössler, E. *Europhys. Lett.* **1995**, *32*, 511.
18. Schönhals, A.; Kremer, F.; Hofmann, A.; Fischer, E. W.; Schlosser, E. *Phys. Rev. Lett.* **1993**, *70*, 3459.
19. Lunkenheimer, P.; Pimenov, A.; Dressel, M.; Goncharov, Y. G.; Böhmer, R.; Loidl, A. *Phys. Rev. Lett.* **1996**, *77*, 318.
20. Schneider, U.; Brand, R.; Lunkenheimer, P.; Loidl, A. *Phys. Rev. Lett.* **2000**, *84*, 5560.

21. Gibson, G. E.; Giauque, W. F. *J. Amer. Chem. Soc.* **1923**, *45*, 93.
22. Parks, G. S.; Huffman, H. M. *J. Phys. Chem.* **1927**, *31*, 1842.
23. Bu, H.S.; Aycock, W.; Cheng, S.Z.D.; Wunderlich,B. *Polymer* **1988**, *29*, 1485.
24. Fujimori, H.; Oguni, M. *J. Chem. Thermodynamics* **1994**, *26*, 367.
25. Hikima, T.; Hanaya, M.; Oguni, M. *Solid State Comm.* **1995**, *93*, 713.
26. Mizukami, M.; Fijimori, H.; Oguni, M. *Prog. Theoretical Phys.* **1997**, *126*, 79.
27. Sillescu, H. *J. Non-Cryst. Solids* **1999**, *243*, 81.
28. Schmidt-Rohr, K.; Spiess, H. W. *Phys. Rev. Lett.* **1991** *66*, 3020.
29. Böhmer, R.; Hinze, G.; Diezemann, G.; Geil, B.; Sillescu, H. *Europhys. Lett.* **1996**, *36*, 55.
30. Böhmer, R.; Hinze, G.; Jorg, T.; Qi, F.; Sillescu, H. *J. Phys. Cond. Mat.* **2000**, *12*, A383.
31. Cicerone, M. T.; Ediger, M. D. *J. Chem. Phys.* **1995**, *103*, 5684.
32. Richert, R. *J Phys. Chem. B* **1997**, *101*, 6323.
33. Böhmer, R.; Chamberlin, R. V.; Diezemann, G.; Geil, B.; Heuer, A.; Hinze, G.; Kuebler, S. C.; Richert, R.; Schiener, B.; Sillescu, H.; Spiess, H. W.; Tracht, U.; Wilhelm, M. *J. Non-Cryst. Solids* **1998**, *235-237*, 1.
34. Wendt, H.; Richert, R. *Phys. Rev. E* **2000**, *61*, 1722.
35. Donth, E. *J. Non-Cryst. Solids* **1982**, *53*, 325; Fischer, E. W.; Donth, E.; Steffen, W. *Phys. Rev. Lett.* **1992**, *68*, 2344.
36. Moynihan, C. T.; Schroeder, J. *J. Non-Cryst. Solids* **1993**, *160*, 52.
37. Mohanty, U. *Adv. Chem. Phys.* **1995**, *89*, 89.
38. Vidal Russell, E.; Walther, E.; Israeloff, N. E.; Alvarez Gomariz, H. *Phys. Rev. Lett.* **1998**, *81*, 1461.
39. Tracht, U.; Wilhelm, M.; Heuer, A.; Feng, H.; Schmidt-Rohr, K.; Spiess, H. W. *Phys. Rev. Lett.* **1998**, *81*, 2727.
40. Vidal Russell, E.; Israeloff, N. E. *Nature* **2000**, *408*, 695.
41. Reinsberg, S. A.; Qiu, X. H.; Wilhelm, M.; Spiess, H. W.; Ediger, M. D. *J. Chem. Phys.* **2001**, *114*, 7299.
42. Hiwatari, Y.; Miyagawa, H.; Odagaki, T. *Sol. St. Ionics* **1991**, *47*, 179.
43. Chamberlin, R. V. *Phys. Rev. B* **1993**, *48*, 15638.
44. Kegel, W.K.; van Blaaderen, A. *Science* **2000**, *287*, 290.
45. Weeks, E.R.; Crocker, J.C.; Levitt, A.C.; Schofield, A.; Weitz, D.A. *Science* **2000**, *287*, 627.
46. Hodge, I.M. *J. Non-Cryst. Solids* **1994**, *169*, 211.
47. Loponen, M.T.; Dynes, R.C.; Narayanamurti, V.; Garno, J.P. *Phys. Rev. B* **1982**, *25*, 1161.
48. De Yoreo, J.J.; Knaak, W.; Meissner, M.; Pohl, R.O. *Phys. Rev. B.* **1986**, *34*, 8828.
49. Sampat, N.; Meissner, M. pg. 105 in *Die Kunst of Phonons*, eds. Paszkiewicz, T. and Rapcewicz, K., Plenum, New York **1994**.
50. Fujimori, H.; Oguni, M. *J. Non-Cryst. Solids* **1994**, *172-174*, 601.
51. Birge, N.O. *Phys. Rev. B* **1986**, *34*, 1631.
52. Dixon, P.K. *Phys. Rev. B* **1990**, *42*, 8179.
53. Christensen, T.; Olsen, N.B. *J. Non-Cryst. Solids* **1998**, *235-237*, 296.
54. Schiener, B.; Böhmer, R.; Loidl, A.; Chamberlin,R.V. *Science* **1996**, *274*, 752.

55. Chamberlin, R. V.; Schiener, B.; Böhmer, R. *Mat. Res. Soc. Symp. Proc.* **1997**, *455*, 117.

56. Leheny, R. L.; Nagel, S. R. *Phys. Rev. B* **1998**, *57*, 5154.

57. Kircher, O.; Schiener, B; Böhmer, R. *Phys. Rev. Lett.* **1998**, *81*, 4520.

58. Richert, R; Böhmer, R. *Phys. Rev. Lett.* **1999**, *83*, 4337.

59. Chamberlin, R.V. *Phys. Rev. Lett.* **1999**, *83*, 5134.

60. Chamberlin, R.V. *Phase Transitions* **1998**, *65*, 169; and to be published.

61. Huang, K. *Statistical Mechanics, 2nd edition*, Wiley, New York, **1987**.

62. Creutz, M. *Phys. Rev. Lett.* **1983**, *50*, 1411; ibid **1992**, *69*, 1002.

63. Creutz, M. *Annals Phys.* **1986**, *167*, 62.

64. Cipra, B. *Science* **2000**, *288*, 1561.

65. Donati, C.; Douglas, J. F.; Kob, W.; Plimpton, S. J.; Poole, P. H.; Glotzer, S. *Phys. Rev. Lett.* **1998**, *80*, 2338.

66. Chamberlin, R. V. to be published.

67. Kubo, R.; Matsuo, K.; Kitahara, K. *J. Stat. Phys.* **1973**, *9*, 51.

68. Hansen, C.; Richert, R.; Fischer, E. W. *J. Non-Cryst. Solids* **1997**, *215*, 293.

69. Anderson, J. E.; Ullmann R. *J. Chem. Phys.* **1967**, *47*, 2178.

70. Wiedersich, J.; Blochowicz, T.; Benkhof, S.; Kudlik, A.; Sorovtsev, N.V.; Tshirwitz, C.; Novikov, V.N.; Rossler, E. *J. Phys. Cond. Mat.* **1999**, *11*, A147.

71. Adam, G.; Gibbs, J. H. *J. Chem. Phys.* **1965**, *43*, 139.

72. Takahara, S.; Yamamuro, O.; Matsuo, T. *J. Phys. Chem.* **1995**, *99*, 9589.

73. Yamamuro, O.; Tsukushi, I.; Lindqvist, A.; Takahara, S.; Ishikawa, M.; Matsuo, T. *J. Phys. Chem.* **1998**, *102*, 1605.

74. Mohanty, U. *J. Chem. Phys.* **1988**, *89*, 3778.

75. Mohanty, U. *J. Chem. Phys.* **1994**, *100*, 5905.

76. Cohen, M. H.; Grest, G. S. *Phys. Rev. B* **1979**, *20*, 1077; and **1981**, *24*, 4091.

77. Anderson, P. W. *Ill-Condensed Matter*, ed. by R. Balian, R. Maynard and G. Toulouse; North-Holland: New York, **1979**; pg. 160.

78. Götze, W.; Sjögren, L. *Rep. Prog. Phys.* **1992**, *55*, 241.

79. Li, G.; Du, W. M.; Sakai, A.; Cummins, H. Z. *Phys. Rev. A* **1992**, *46*, 3343.

80. Lunkenheimer, P.; Pimenov, A.; Dressel, M.; Schiener, B.; Schneider, U.; Loidl, A. *Prog. Theor. Physics. Suppl.* **1997**, *126*, 123.

81. Wuttke, J.; Hernandez, J.; Li, G.; Coddens, G.; Cummins, H. Z.; Fujara, F.; Petry, W.; Sillescu, H. *Phys. Rev. Lett.* **1994**, *72*, 3052.

82. Ye, J. Y.; Hattori, T.; Nakatsuka, H.; Murayama, Y.; Ishikawa, M. *Phys. Rev. B* **1997**, *56*, 5286.

Chapter 18

On a Connection between Replica Symmetry Breaking and Narayanaswamy–Gardon Nonlinear Parameters

Udayan Mohanty

Eugene F. Merkert Chemistry Center, Boston College, 140 Commonwealth Avenue, Chestnut Hill, MA 02467

A connection is established between Parisi's replica symmetry breaking parameter, $m_{replica}(T)$, that indicates whether the various minima in configurational space are as distinct as imaginable, and the experimental measurable Narayanaswamy-Gardon non-linear parameter x that characterizes the deviation of a glass-forming liquid from equilibrium. The linear temperature dependence of $m_{replica}(T)$ is in agreement with one step replica symmetry breaking. The slope of $m_{replica}(T)$ is governed by the non-linearity of the supercooled state.

1. Introduction

Supercooled liquids evolve in a restricted part of the configuration space. At or near the glass transition temperature, the configurational space separates into a large number of regions that are mutually inaccessible (1). To describe the low temperature behavior of glassy systems, Parisi and coworkers have introduced an order-parameter that measures the extent by which different regions of the configuration space spanned by the supercooled and the glassy states of matter are similar (1). They introduce an order parameter $q(z,y)$ that measures the overlap of the two replicas z and y (1). High values of q imply that z and y have similar configurations, while low values of q imply that the configurations are significantly different (1). Non-zero Boltzmann-weight is assigned to those configurations which are at a specified distance away from configuration y.

In model system, Parisi and coworkers have demonstrated that for temperatures below the Kauzmann temperature, replica symmetry is broken and the parameter that characterizes this phenomenon is linear in temperature (1)

$$m_{replica}(T) = \frac{T}{T_k}.$$ (1)

The various minima are as distinct as imaginable, and there is no correlation between molecules in the two minima (1). To put it differently, two configurations are orthogonal if they are not in the same minimum.

The non-linearity of the glassy state is generally described by the Narayanaswamy-Gardon relaxation time relation relation (2)

$$\tau = \tau_o \, exp(\frac{x\Delta h^*}{RT} + (1-x)\frac{\Delta h^*}{RT_f}),$$ (2)

where x is the non-linear parameter that has been experimentally measured for glass forming liquids, Δh^* is the enthalpy of activation near the glass transition temperature T_g, and T_f is the fictive temperature, i.e., the temperature at which the system would be at equilibrium (2). For equilibrium liquid, $T_f = T$, while in the glassy state, $T_f = T_g$.

We now demonstrate a connection between the replica symmetry breaking parameter $m_{replica}(T)$, and the experimental measurable Narayanaswamy-Gardon non-linear parameter x that characterizes the deviation of the system from equilibrium. The paper is organized as follows. In Section 1, we establish some relations between the fragility index and experimentally measurable properties of supercooled liquids. In section 2, a link is established between the Narayanaswamy-Gardon non-linear parameter that serves as a measure of how far the system is away from equilibrium and the replica symmetry breaking parameter that indicates whether the various minima are as distinct as imaginable.

2. Fragility index

The starting point of the investigation is the Adam-Gibbs model for the transition probability $w(T)$ of cooperative rearrangements in a glass forming liquid in an isothermal-isobaric ensemble (3)

$$w(T) = A \, exp-(z\Delta\mu / k_B T). \tag{3}$$

z is the critical number of molecules that allows cooperative rearrangements, k_B is the Boltzmann constant, T is the absolute temperature, the constant A is weakly temperature dependent, and $\Delta\mu$ is the change in chemical potential per molecule due to cooperative rearrangements (3). The critical number of molecules is inversely proportional to the configurational entropy of the supercooled liquid (3). A generalization of the Adam-Gibbs model has been developed based on renormalization group approach (4,5). Such an analysis lead to an explicit relation between the critical size that allows cooperative rearrangements and the total entropy $S(T)$ of the liquid (4,5)

$$z \approx \frac{N_A s^*}{S(T)}. \tag{4}$$

Here, N_A is the Avogardo number, and s^* is the entropy of the cooperatively rearranging region.

A measure of the characteristics of strong and fragile liquids is the deviation of the temperature dependence of the relaxation time $\tau(T)$ from Arrhenius like behavior as measured by the fragility index m (6)

$$m = \frac{d \, log\tau(T)}{d(T_g / T)}\bigg|_{T=T_g}. \tag{5}$$

The minimum value of the fragility index $m_{min} = log(\tau(T_g)/\tau_o)$, where τ_o is usually identified with vibration lifetime of phonons in high temperature limit $\tau_o = 10^{-14} s$ (6).

To provide insights into the concept of fragility index, we have considered a generalization of Gibbs-DiMarzio configuration entropy model of glasses that includes changes in lattice vibrations due to temperature dependence of the force constants (7,8). After tedious algebra, a relation between the fragility index, heat capacity and the entropy of the polymeric liquid is obtained (9)

252

$$\frac{m}{m_{min}} = 1 + \frac{\Delta C_p(T_g)}{\Delta S(T_g)}.$$
(6)

The relation can also be obtained on combining Eqs. (3)-(5), i.e., without assuming the temperature dependence of the heat capacity as given by the DiMarzio-Dowell configurational entropy model (8).

The activation energy in the liquid and the glass states are given, respectively, by

$$\frac{d\log\tau}{d(1/T)}\bigg|_{liquid} = \frac{\Delta\mu z}{k_B} - \frac{\Delta\mu z}{k_B}(\frac{\partial z}{\partial T})_P = \frac{\Delta h^*_{liquid}}{R},$$

$$\frac{d\log\tau}{d(1/T)}\bigg|_{glass} = \frac{\Delta\mu z}{k_B} = \frac{\Delta h^*_{glass}}{R}.$$
(7)

The ratio of the two activation energies yield the Narayanaswamy-Gardon non-linear parameter x (10)

$$x = \frac{1}{1 - \frac{T}{z}(\frac{\partial z}{\partial T})_P}.$$
(8a)

But the fragility can be shown to be given by the relation (10)

$$m/m_{min} = 1 - \frac{T}{z}(\frac{\partial z}{\partial T})_P,$$
(8b)

and therefore inversely related to the non-linear parameter, $m/m_{min} = 1/x$ (10).

3. Replica symmetry breaking and Narayanaswamy-Gardon non-linear parameters

We now demonstrate a connection between the replica symmetry breaking parameter $m_{replica}(T)$, and the experimental measurable Narayanaswamy-Gardon non-linear parameter x that characterizes the deviation of the system from equilibrium.

The starting point of the analysis is the partition function expressed in the inherent structure viewpoint (1,11) at low temperatures for a polymeric liquid at some fictive temperature T_f — the temperature at which the system would be at equilibrium

$$e^{-\beta_f N\Phi} = \sum_\varphi e^{-\beta_f N\varphi} = \int_{f_{min}}^{f_{max}} df e^{-N(\beta_f f - s_c(f,T_f))},$$

(9)

where Φ is the total free energy per monomer, N is the number of monomers, f is the free energy per monomer, $\beta_f = 1/k_B T_f$, and the sum is over the distinct cells or minima. Define $\Phi(f) = f - s_c(f,T_f)/\beta_f$; by saddle point method, $\Phi(f) = min_f(f^* - k_B T_f s_c(f^*,T_f))$, where f^* is the value of f that minimizes $\Phi(f)$ provided f^* lies inside the interval $f_{min} < f < f_{max}$ (11,7). The other possibility, first realized by Gibbs-DiMarzio, is when the minimum is on the boundary, i.e., the saddle point is attached to f_{min}; in this case $\Phi = f_{min}$ (11,7).

Monasson has argued that by constraining m replicas in the same state, one can construct the configurational entropy (12). If $m_{replica}$ replicas are constrained to be in the same state, then the minimum value of f satisfies (1)

$$m_{replica} = k_B T_f (\frac{\partial s_c/k_B}{\partial f}).$$

(10)

For temperatures below the glass transition temperature, we rewrite the Narayanaswamy-Gardon non-linear parameter x as

$$x = \cfrac{1}{1 + \cfrac{T_g}{s_c(T_f) + s_{vib}(T_f)}(\frac{\partial s_c}{\partial f})T_f(\frac{\partial f}{\partial T})T_f + \cfrac{T_g}{s_c(T_f) + s_{vib}(T_f)}(\frac{\partial s_{vib}}{\partial f})T_f}$$

(11)

where the entropy per particle $s = S/N$ has been decomposed in terms of configuration and vibration contributions $s_c + s_{vib}$. On imposing the condition that if $m_{replica}$ replicas are constrained to be in the same state, then the minimum value of f satisfies Eq.(10), and assuming, in accordance with the

Gibbs-DiMarzio model (8), that for temperatures less than the Kauzmann temperature, the contribution to the entropy is from the vibration degrees of freedom, one obtains

$$m_{replica} = 1.28 \frac{T_f}{T_k} (\frac{1}{x} - 1) - \frac{c_{vib}}{s_{vib}}. \tag{12}$$

Several comments are in order regarding Eq. (12), which is the central result of this paper. First, in deriving Eq. (12), one has approximated $s(T_f)/(\frac{\partial f}{\partial T})T_f$ by unity and utilized a result shown elsewhere that the ratio of $T_g/T_o = 1.28$ for over dozen glass-forming liquids (5). This result has been justified without recourse to Adam-Gibbs configurational entropy arguments (5). Second, the linear temperature dependence of $m_{replica}$ is in agreement with one step replica symmetry breaking prediction by Parisi and coworkers (1). Third, the slope of $m_{replica}(T)$ is governed by the non-linearity of the supercooled state. Fourth, the second term on the right hand side of Eq. (12) is small but non-zero at finite T.

Acknowledgments

I thank Giorgio Parisi for penetrating discussion on replica symmetry breaking formalism.

References

1. (a) Parisi. G. in Supercooled Liquids: Advances and Novel Applications, Eds. Fourkas, J. T.; Kivelson, D.; Mohanty, U.; Nelson, K. A. *ACS Symposium Series* **1997**, 676, 110-121; (b) Franz, S.; Parisi, G. Preprint, Dec. **1997**; cond-mat/9701033. (c) Franz, S.; Parisi, G. Preprint, Dec. **1997**, pages 1-18; cond-mat/9711215. (d) Mezard, M.; and Parisi, G. Preprint, Cond-mat/9812180. (e) Cardenas, M.; Franz, S.; Parisi, G. Preprint, Dec. **1997**, pages 1-23.
2. Hodge, I. M. *J. Non.Cryst. Solids* **1994**, 169, 211-266.
3. Adam. G.; Gibbs, J. H. *J. Chem. Phys.* **1965**, 43, 139-146.
4. Mohanty, U. in Supercooled Liquids: Advances and Novel Applications; Fourkas, J. T.; Kivelson, D.; Mohanty, U.; Nelson, K. A. Ed.; *ACS Symposium Series* **1997**, 676, 95-109.

5. Mohanty, U. *Physica* **1991**, A 177, 345-355.
6. Hodge, I. M. *J. Non.Cryst. Solids* **1996**, 202, 164-172.
7. Gibbs, J. H.; DiMarzio, E. A. *J. Chem. Phys.* **1958**, 28, 373.
8. DiMarzio, E. A.; Dowell, F. *J. Appl. Phys.* **1979**, 50, 6061-6066.
9. (a) Mohanty, U.; N. Craig, N.; Fourkas, J, T. *J. Chem.Phys.* **2001** (in press).
    (b) Mohanty, U.; N. Craig, N.; Fourkas, J, T. *Phys. Rev. E.* **2001** (in press).
10. Mohanty, U. unpublished, **2000**.
11. (a) Montero, M. J. R.; Mohanty, U.; Brey, J. *J. Chem Phys.* **1993**, 33, 9979-9983. (b) Mohanty, U. *Adv. Chem. Phys.* **1994**, LXXXIX, 89.
12. Monasson, R. *Phys. Rev. Lett.* **1995**, 75, 2847.

Chapter 19

Intramolecular Motions in Simple Glass-Forming Liquids Studied by Deuteron NMR

H. Sillescu, R. Böhmer, A. Döß, G. Hinze, Th. Jörg, and F. Qi

Institut für Physikalische Chemie, Johannes Gutenberg-Universität, Jakob-Weldev Weg 15, D–55099 Mainz, Germany

In most glass-forming liquids there are intramolecular motions that may interfere with secondary relaxation processes considered to be signatures of the glass transition, e.g., the Johari-Goldstein Process. To explore these motions we have investigated deuteron spin-lattice relaxation, spectral line shapes, and the response to special spin-echo pulse sequences in partially deuterated samples of typical models for van der Waals liquids, namely, ortho-terphenyl, salol, and propylene carbonate. Our results yield information on librational motion of the outer phenyl groups in ortho-terphenyl, on intra-molecular phenyl group rotation in salol, and methyl group rotation in propylene carbonate. These motions are often neglected in dynamical studies of neutron and light scattering, or dielectric relaxation and NMR. We discuss their influence on relaxations close to the glass transition.

Introduction

In attempts to identify universal dynamics at the glass transition there was always the problem of excluding intramolecular motions. In principle, one should work with glass-forming liquids consisting of completely rigid molecules. Indeed, in their pioneering dielectric experiments Johari and Goldstein were able to show that a β-process is observable in solutions of rigid polar molecules in glass-forming liquids (1). This Johari-Goldstein (JG) β-process was also observed in several pure liquids in their glass state and sometimes even in some region above the glass transition temperature, T_g (2). However, most liquids consisting of rigid molecules have a strong tendency to crystallize or they have other disadvantages such as chemical instability or very low T_g's that prevent investigation by different relaxation and scattering techniques each having its own experimental limitations. For this reason, the most extensively studied glass-forming liquids consist of flexible molecules where intramolecular motions may or may not interfere with "universal" dynamics. We have chosen some systems listed in Fig.1 where intramolecular motions could be studied by deuteron NMR techniques. Excluded are systems in which the formation and dissociation of *inter*molecular hydrogen bonds can occur on the same time scale as α- or β-processes. In ²H-NMR studies of partially deuterated glyceroles no effects of intramolecular motions were observed (3) and they are certainly negligible in comparison with effects of H-bonding dynamics.

Intramolecular motions play a minor role in studies of the α-process, except, if its "high frequency wing" (4) is investigated. However, they can interfere with the "slow" JG-β-process as well as with the "fast" β-process addressed in experimental studies related with predictions of mode coupling theory (MCT) (5). Here, orthoterphenyl (OTP), salol, and propylene carbonate are the most extensively studied fragile glass-forming liquids. However, already the first neutron scattering study looking for a MCT β-process in a fragile glass-former was on trinaphthyl benzene (TNB) (6). Here the librational motions of the naphthyl groups around their bonds to the central benzene ring increase in amplitude and become full rotations on approaching the liquid state from lower temperatures. Therefore, the motion of the proton scattering centers reflect these intramolecular motions in addition to the influence of density fluctuations treated by MCT. In a ²H-NMR study of TNB (7) the spin-lattice relaxation decay functions could be fitted by a superposition of two exponentials. The relaxation times, $T_1^{(int)}$ and $T_1^{(iso)}$, obtained from this fit refer to 14 deuterons affected ($T_1^{(int)}$) and 10 deuterons not affected ($T_1^{(iso)}$) by the internal rotation of naphthyl groups. The corresponding correlation times, τ_{int} and τ_{iso}, differ by a factor of ~10 in the liquid regime. In the glass the influence of intramolecular motion could not be identified. This is possibly because the librational amplitudes here are too small. This

258

conjecture is in accord with the small proton mean square displacements accessible from the Debye-Waller factor determined by neutron scattering (6).

In the following sections, the results of our NMR studies of partially deuterated orthoterphenyl(8), salol(9), toluene(10) and propylene carbonate(10,11) are summarized and discussed with particular emphasis on the influence of intramolecular motions in comparison with other relaxation and scattering results.

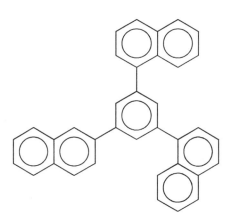

TNB (T$_g$ = 342K)

OTP (T$_g$ = 243K)

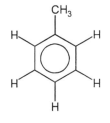

toluene (T$_g$ = 117K)

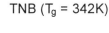

propylene carbonate
(T$_g$ = 157K)

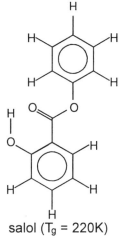

salol (T$_g$ = 220K)

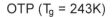

Figure 1 . Structure and glass transition temperatures, T$_g$, of the glass-forming liquids discussed in the Text. TNB stands for ααβ-trinaphthylbenzene (12) and OTP for orthoterphenyl.

Orthoterphenyl

Orthoterphenyl (OTP) is the most extensively studied glass-forming van der Waals liquid. In particular, detailed comparisons of neutron scattering results with MCT predictions were carried out (*13*). Here, possible librational motions of the lateral phenyl groups should be on the same time scale as the fast β-process of MCT. However, neutron scattering measurements of the Debye-Waller factor (normalized to its value at 180 K, performed in a Q-range of 1.2 - 4.8 Å^{-1} at 8 μeV energy resolution) showed no observable difference in a T-range of 193 - 284 K between normal and fully deuterated OTP and OTP-d$_{10}$ deuterated at the lateral phenyl-rings (*14*). It was concluded "that any intra-molecular phenyl-ring motions can be safely excluded as dominant dynamic mechanisms for the glass transition β-process" (*14*).

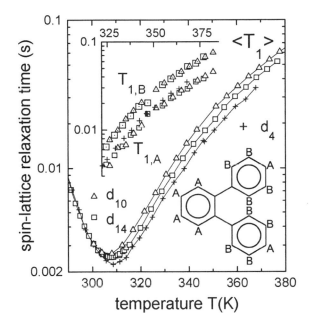

Figure 2. Average ^2H spin-lattice relaxation times $\langle T_1 \rangle$ of OTP-d$_4$, OTP-d$_{10}$, and OTP-d$_{14}$. The lower inset shows OTP with B indicating those deuterons that are affected by torsional phenyl-ring motions whereas those labeled by A are only sensitive to whole molecule reorientation. The upper inset shows the relaxation times $T_{1,A}$ and $T_{1,B}$ obtained from a biexponential fit.
(Adapted with permission from reference 8. Copyright 2000 EDP Sciences.)

After OTP-d_4 , deuterated at the benzene ring, had become available to us, we explored the possibility of phenyl-ring motions by ^{2}H-NMR (8). In Figure 2, we show the average spin-lattice relaxation times of OTP-d_4 , OTP-d_{10} , and OTP-d_{14} as well as the values $T_{1,A}$ and $T_{1,B}$ obtained from a biexponential fit of the spin-magnetization decay curves (8). The results are similar to those discussed above for TNB, however, the data are more accurate and it is reassuring that $T_{1,A}$ agrees with T_1 of OTP-d_4 with no deuterons at the phenyl-rings.

A quantitative interpretation of the difference between $T_{1,A}$ and $T_{1,B}$ is possible in terms of special models for the intramolecular motions. Here the maximum angular range of the phenyl-ring librations as well as the apparent motional enhancement depend upon the model assumptions (15). The ratio $T_{1,A}/T_{1,B}$ = 0.6 ± 0.05 found at $T \geq 325$ K is compatible with rather large angles ($\sim 60°$). The time scale of the librational motions is in the ns to ps range, comparable with that probed by neutron scattering at much lower temperatures, $T \leq 284$ K (14) where, however, the librational amplitudes should be much smaller. We could also detect no additional contribution of phenyl-ring motion at $T < 300$ K with ^{2}H-NMR experiments (see ref. (8) for further details) which thus support our conclusion drawn from neutron scattering.

Salol

There are many experimental studies of salol (phenyl salicylate) by dielectric spectroscopy (16), and various light scattering (17,18) and optical Kerr effect (19,20) methods since the molecular electric dipole moment and polarizability as well as T_g = 220 K are in ranges favorable for these techniques. The internal flexibility of the salol molecule that is apparent from its structure (see Figure 1) was rarely addressed in these studies, but is at the focus of our recent ^{13}C and ^{2}H NMR study(9). The ^{13}C nuclei in CH-bonds of salol are well separated by their different chemical shifts in the NMR spectrum at $T > 290$ K and due to their equal ^{13}C-^{1}H dipolar couplings the rotational correlation times can be easily determined from the T_1-values. Three different correlation times, $\tau_\perp$, $\tau_\parallel$, and τ_{int} could be distinguished from the ^{13}C spins in p-positions, salicyl-ring-o-positions, and phenyl-ring-o-positions, respectively. From the fact that both CH-bonds in p-positions have the same $\tau_\perp$ we conclude that no phenyl group rotation around the central (ester) C - O axis is detectable. However, since the ratio $\tau_\perp /$ $\tau_{int} \simeq 6.5$ exceeds that of $\tau_\perp / \tau_\parallel \simeq 3$ there must be an internal motion of the phenyl-ring around its axis to the central O-atom in addition to the anisotropy of rotational diffusion detected from the difference of $\tau_\parallel$ and $\tau_\perp$. This additional motion was also analyzed by ^{2}H-T_1 of phenyl-ring deuterated salol-d_5 . The results agree nicely with those of the ^{13}C-NMR analysis though the details are somewhat different since distributions of correlation times were included in the ^{2}H-T_1 evaluation which yields $\langle \tau_\perp \rangle / \langle \tau_{\text{int}} \rangle \simeq 4$ (9).

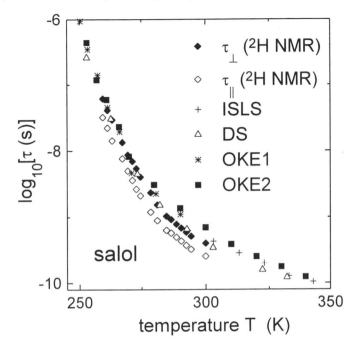

Figure 3. Rotational correlation times in salol. The NMR data are taken from ref. (9). Also included are data from impulsive stimulated light scattering (ISLS, ref. 18) dielectric spectroscopy (DS, ref. 21) and optical Kerr-effect (OKE) techniques, see refs. (19,20).

The correlation times from our ^{2}H-NMR experiments are included in Figure 3 where we also show results from dielectric spectroscopy (21), impulsive stimulated light scattering (ISLS) (18), and the optical Kerr effect (OKE) (19,20). The good agreement of the dielectric relaxation times, τ_{diel}, with $\langle \tau_\perp \rangle$ can be explained as resulting from the balance of two opposing effects. First, τ_{diel} should be a superposition of $\tau_\perp$ and $\tau_\parallel$ since the dipole orientation in the molecule changes also by the faster rotations around the long molecular axis. Second, correlation times for dipole reorientation ($l=1$) can exceed those determined by NMR ($l=2$) by up to a factor of 3 in the limit of small angular step rotational diffusion(22). The larger values of the light scattering and Kerr effect correlation times appear surprising at first. However, the difference can be qualitatively understood from the different averaging inherent in the experimental data analysis. The τ_{OKE} ($=\tau_\alpha$) -values from the decay of OKE intensities were obtained from a fit of only the long time regime (t $\gtrsim \tau_\alpha$) whereas $\langle\tau_\perp\rangle$ averages also over the medium and short time regime of a "distribution of correlation times"(22). Since neither the motional anisotropy ($\tau_\perp / \tau_\parallel$) nor the intramolecular phenyl-ring motion (τ_{int}) was considered in the MCT analysis of τ_α the good agreement (T_c = 253 K) may be fortuitous to some extent. However, the relatively small difference between $\langle\tau_\perp\rangle$ and $\langle\tau_{int}\rangle$ can probably be ignored in comparison with the consequences of the assumption that the rotational degrees of freedom reflect the dynamics of the density fluctuations considered in the analysis by idealized MCT(5,23).

Methyl group rotation in toluene and propylene carbonate

Various aspects of methyl group rotation have been studied in many glass-forming polymers(24) and liquids(25) as they can exhibit rather different phenomena ranging from quantum tunneling motion (26) to slow classical barrier hopping depending upon the size of the intermolecular potential. In Figure 4 we show ^2H spin-lattice relaxation times of the two very fragile van der Waals liquids toluene and propylene carbonate where CH_3 rotation is the only intramolecular motion. By comparing the T_l -values of the ring deuterated with the methyl group deuterated compounds it becomes obvious that CH_3 rotation is effectively decoupled in toluene. In contrast, the much smaller decoupling in propylene carbonate is no longer detectable at higher temperatures. In both systems T_l can be quantitatively described by the Woessner model(27) [see fit curves in Figure 4(10)] treating molecular reorientation and CH_3 rotation as independent processes which are characterized by appropriate distributions of the two correlation times τ_R and τ_{int} , respectively. For τ_R we have chosen the Cole-Davidson distribution (22) [similar to the Kohlrausch-Williams-Watts distribution (10)] with a T independent width of about 2 decades in time. The T dependence of the average, $\langle\tau_R\rangle$, agrees with that of the shear viscosity and relaxation times of other experiments which probe the overall molecular reorientation. [For details, see refs. (10,11).]

The T dependence of CH_3 rotation can be described by the Arrhenius equation, $\tau_{int} = \tau_0 \exp(E/T)$, where the energy barrier, E , is distributed over a range of ~ 100-500 K in toluene whereas a constant value was used for the fit of T_l of propylene carbonate in Figure 4(10). The broad E distribution in toluene causes a τ_{int} distribution that extends over ~4 decades in the low T range. Here τ_{int} is of the order of the inverse Larmor frequency, ω_L^{-1} , at the T_l minimum and the averages $\langle T_1 \rangle$ and $\langle T_1^{-1}\rangle^{-1}$ differ dramatically (10). For propylene carbonate, on the other hand, the τ_{int} distribution narrows to a width below one decade at high

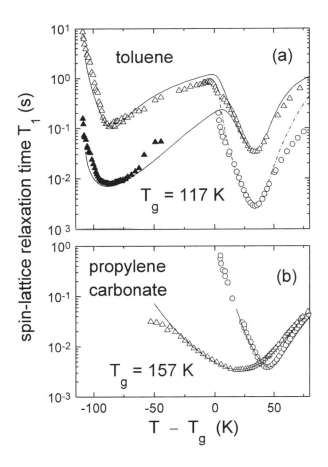

Figure 4. ^2H *spin-lattice relaxation times, T_1 : (a) ring deuterated toluene-d_5 and methyl group deuterated toluene-d_3 . (b) ring deuterated propylene carbonate-d_3, and methyl group deuterated propylene carbonate-d_3 . Circles represent data for the ring deuterated substances, triangles those for the methyl deuterated ones. Open symbols correspond to time averages, full symbols to rate averages. See text and ref. (10) for explanation of fitting curves.*

T which explains why no E distribution was assumed here [(*10*), see however: Qi et al., to be published].

Propylene carbonate has been investigated by different relaxation and scattering techniques, in particular, at temperatures, $T \gtrsim T_c = 182$ K, which was identified as the MCT critical temperature in a recent analysis of neutron scattering and dynamic light scattering (28). The analysis of incoherent neutron scattering was based on data determined at temperatures between 210 K and 260 K at frequencies, $\omega/2\pi$, in a range of ~ 1 - 500 GHz. It is apparent from the T_1 values in Figure 4 that CH_3 rotation is still faster than the overall molecular reorientation at 210 K whereas no difference is detectable at 260 K. Since the neutron scattering experiments were done with fully protonated propylene carbonate (29) one should expect a contribution originating from CH_3 rotation at the low temperature which is absent in the light scattering and dielectric results since the molecular polarizability and dipole moment is hardly affected by CH_3 rotation. Since the frequency, ω_σ, of the susceptibility minimum is lower in the neutron scattering than in the light scattering results the difference will become larger if the influence of CH_3 rotation is subtracted. It is difficult to give a quantitative estimate of this effect from the T_1 values in Figure 4. However, one should note that T_1 is dominated by the low frequency side of the spectral density whereas the neutron scattering analysis is done at frequencies around ω_σ above the position of the α-relaxation peak. Therefore, the perturbation by CH_3 rotation should be smaller than in the case of ^2H- T_1. We should also mention that the different experimental results from dielectric spectroscopy, light scattering and neutron scattering were addressed in a recent theoretical study where propylene carbonate was described by a two-component schematic model of the MCT (30)

Conclusions

We have shown that in ^2H-NMR studies of partially deuterated glass-forming liquids intramolecular motions can be identified in the glass transition region which possibly interfere with processes considered as "universal". In particular, one should be careful in neutron scattering experiments where the motion of the proton scattering centers may be partly due to intramolecular motions. This was investigated in detail in orthoterphenyl and propylene carbonate where the effects of intramolecular motions where shown to be small in comparison with other non universal effects.

The examples chosen (Figure 1) all have a very high fragility (i.e., a large T dependence of the α-process close to T_g) and they show no indication of a Johari-Goldstein (JG) β-process(4,16), except for toluene where, however, the CH_3 rotation occurs on a very fast time scale in comparison of that of the JG process. We should also note that the amplitude and time scale of intramolecular motions changes much less with temperature than those of the JG process which merges with the α-process at temperatures close to $T_c > T_g$ (2). Finally, we should mention that all NMR experiments were done in well annealed samples where a particular process that was recently observed in quenched orthoterphenyl and salol (16) is not detectable.

Acknowledgment

We wish to thank Herbert Zimmermann for his permanent willingness to prepare the partially deuterated substances we wanted to investigate. Support by the Deutsche Forschungsgemeinschaft (SFB 262) is gratefully acknowledged.

References

1. Johari,G.P.; Goldstein,M. *J. Chem. Phys.* **1970,***53*, 2372.
2. Wagner,H.; Richert,R. *J.Phys.Chem.B* **1999,***103,*4071.
3. Schnauss,W.;Fujara,F.;Sillescu,H. *J.Chem.Phys* 1992, *97* ,1378, and references therein.
4. Blochowitz,T.; Kudlik,A.; Benkhof,S.; Senker,J.; Rössler,E.; Hinze,G. *J.Chem.Phys.***1999,***110,*12011, and references therein.
5. Götze,W.; Sjögren,L. *Repts.Progr.Phys.* **1992,***55,*241. Götze,W. *J.Phys.: Cond.Matt.***1999,***11,*A1.
6. Fujara,F.; Petry,W. *Europhys.Lett.***1987,***4,*928.
7. Schnauss, *PhD-Diss.*, Univ.Mainz 1991.
8. Jörg,T.; Böhmer,R.; Sillescu,H.; Zimmermann,H. *Europhys.Lett.* **2000,***49,*748.
9. Döß,A.; Hinze,G.; Böhmer,R.; Sillescu,H.; Kolshorn,H.; Vogel,M.; Zimmermann,H. *J.Chem.Phys.* **2000,** *112,*5884.
10. Qi,F.; Hinze,G.; Böhmer,R.; Sillescu,H.; Zimmermann,H., Chem. Phys. Lett. (accepted)
11. Qi,F.; Schug,K.U.; Döß,A.; Dupont,S.; Böhmer,R.; Sillescu,H.; Kolshorn,H.; Zimmermann,H.; *J. Chem. Phys.* **2000,***112,* 9455.
12. It was shown by C.M.Whitaker and R.J.McMahon (*J.Phys.Chem.* **1996,***100,*1081) that this substance rather than tri-α-naphthylbenzene was investigated in refs.(*3,6,7*).
13. Tölle,A.;Schober,H.; Wuttke.;Fujara,F.*Phys. Rev. E* **1997,***56,*809, and references therein
14. Debus,O.; Zimmermann,H.; Bartsch,E.; Fujara,F.; Kiebel,M.; Petry,W.; Sillescu,H. *Chemical Physics Letters* 1991 *,180,* 271.
15. Witteborg,R.J.; Szabo,A. *J.Chem.Phys.***1978,***69,*1722, and references therein.
16. Wagner,H.; Richert,R. *J. Chem. Phys.* **1999,***110,*11660, and references therein.

266

17 . Li,G.; Du,W.M.; Sakai,A.; Cummins,H.Z. *Phys.Rev.A* **1992**,*46*,3343.
18 . Yang,Y.; Nelson,K.A.; *J. Chem. Phys.* **1995**,*103*,7732.
19 . Hinze,G.; Francis,R.; Fayer,M.D.; *J.Chem.Phys.* **1999**,*111*,2710.
20 . Hinze,G.; Brace,D.D.; Gottke,S.D.; Fayer,M.D.; *J.Chem.Phys.***2000,**
       accepted.
21 . Schönhals,A.; Kremer,F.; Hofmann,A.; Fischer,E.W.; Schlosser,E. *Phys. Rev.Lett.***1993,***70*,3459.
22 . Böttcher, C.F.J.; Bordewijk, P. *Theory of Electric Polarization, Vol II*, Elsevier: Amsterdam 1978
23 . Schilling,R.; Scheidsteger,T.*Phys.Rev.E* **1997**,*56*,2932.
24 . O'Connor, R. D.; Ginsburg, E. J.; Blum, F. D. *J. Chem. Phys.* **2000**,*112*, 7247, and references therein.
25 . Hertz,H.G.; *Progr. NMR Spectrosc.* **1983**,*16*, 115, and references therein.
26 . Börner,K.; Diezemann,G.; Rössler,E.; Vieth,H.M.; *Chem. Phys. Lett.* **1991**,*181*, 563, and references therein.
27 . Woessner,D.E.; *J. Chem. Phys.* **1961**,*36*, 1; *Adv. Molec. Relax. Proc.* **1972**,*3*, 181.
28 . Wuttke,J.: Ohl,M.; Goldammer,M.; Roth,S.: Lunkenheimer,P.; Kahn,R.; Rufflé,B.; Lechner,R.; Berg,M.A.; *Phys. Rev. E* 2000,*61*, 2730, and references cited therein.
29 . Work with methyl group deuterated propylene carbonate is in progress. (Wuttke,J.; private communication)
30 . Götze,W.; Voigtmann,Th. *Phys.Rev.E* **2000**,*61*,4133,

Confined Liquids

Chapter 20

Molecular Dynamics in Confining Geometries

A. Huwe and F. Kremer[*]

Department of Physics, University of Leipzig, D–04103 Leipzig, Germany

Broadband dielectric studies (10^{-3} Hz - 10^9 Hz) on the
molecular dynamics of low molecular weight glass forming
liquids in confining geometries are reviewed. In these systems an interplay between surface- and confinement effects
is observed: Due to the interaction of the guest molecules
with the host system their dynamics is slowed down and
hence the glass transition temperature is increased. This
effect can be counterbalanced by lubricating the inner surfaces of the pores causing the dynamics of the molecules inside the pores to decouple from the solid walls. – Confinement effects cause the molecular dynamics of an embedded
liquid to be orders of magnitude faster than the bulk liquid. The glass transition temperature is also observed to
dramatically decrease with the strength of confinement. It
reflects the inherent length scale of the dynamic glass transition of the embedded liquid which increases with decreasing temperature. If it becomes comparable with the size of
the confining space the molecular dynamics changes from
a Vogel-Fulcher-Tammann- to an Arrhenius-type temperature dependence. This, in turn, proves the cooperative
nature of the dynamic glass transition.

© 2002 American Chemical Society

Introduction

The molecular dynamics of liquids, supercooled liquids and glasses is an interesting area of present research and discussion [1–10]. Studying the molecular dynamics of glass forming liquids in confining geometries may help to test predictions of models and concepts concerning the dynamic glass transition. One important question is the length scale on which cooperative motions of a liquid take place [11–20].

Recent experiments turn out, that this length scale is much larger for van-der-Waals liquids than for H-bond forming systems. Nanoporous sol-gel glasses with pore diameters of 2.5, 5.0 and 7.5 nm have shown to be a favourable host system to measure the moelcular dynamics of the "quasi"-van-der-Waals liquid salol in confining geometries [10,19,20]. In contrast for H-bond forming systems like propylene and ethylene glycol nanopores with much smaller diameter have to be used to study confinement effects [9,10,18,21,22].

Modern solid state chemistry enables to realize the ideal of a single isolated molecule in one zeolitic cage [23]. Varying the size and the topology of the zeolitic host system allows to examine the transition from a single-molecule dynamics to that of a bulk liquid. It is evident that this problem is related to the understanding of the cooperative dynamics and its inherent length scale in glass forming liquids [7–17,24].

Experimental

Nanoporous glasses that have been used so far as host materials for the study of confinement effects have the disadvantages of possessing broad pore-size distributions and an unknown topology of the pores. A new class of silica materials, named M41S, avoids these drawbacks. The pore-size distribution of these materials is narrow and the mean pore size can be adjusted between 2 to 10 nm [25–27]. Several different materials with well-defined topologies exist and can be prepared reproducibly [25–28]. MCM-41, for example, possesses parallel uni-dimensional channels which are arranged in a hexagonal array. MCM-48 contains three dimensionally connected pores (see Figure 1).

M41S materials can therefore be expected to perform excellently as host systems for further investigations of the dynamics in confined geometries. To realize a spatial confinement on a length scale below 1 nm zeolites are suitable materials. Varying the size and the topology of the zeolitic systems allows to study the transition from the dynamics of an isolated molecule to that of a bulk liquid.

It is possible to realize one single molecule in a zeolitic cage. This is done by use of a structure-directed synthesis. The structure-directing agents, for example ethylene glycol (EG), which control the formation of silica sodalite [23,29], become occluded during synthesis and cannot escape from the cages. Zeolite beta, silicalite, ZSM-5 and $AlPO_4$-5 have pore systems consisting of channels .

Zeolite beta, an aluminosilicate, has two types of channels. In [100]- and [010]-directions they have an elliptical cross-section of 0.76 nm x 0.64 nm. The channels in the [001]-directions have a diameter of 0.55 nm [30]. Silicalite and ZSM-5 have the same structure. They have two types of elliptical channels forming a three-dimensional pore system. Their cross-sections are 0.56 nm x 0.53 nm and 0.55 nm x 0.51 nm [31]. Silicalite consists of pure SiO_2 whereas ZSM-5 is an aluminosilicate. To maintain the neutral charge of the zeolite ZSM-5 contains counter ions in the channels. Our sample has a Si:Al ratio of 25 and is named H-ZSM-5 to indicate that the counter ions are protons. The zeolite beta sample which was used contains protons as well and it has a Si:Al ratio of 40. In contrast, $AlPO_4$-5 is an aluminophosphate with a one-dimensional pore system. The channels are arranged in a hexagonal array and have a diameter of 0.73 nm which is comparable to the size of the channels in zeolite beta.

All MCM-materials and zeolites are provided as small (< 0.1mm) crystalites which have to be filled from the vapor phase. For some glass forming liquids like salol this procedure is not practicable as they decompose at elevated temperatures (400 K). Therefore additional studies using nanoporous sol-gel glasses from Geltech Inc., USA with specific pore sizes of 2.5 nm, 5.0 nm, and 7.5 nm and a narrow pore size distribution were performed. The preparation of the samples and the details of the dielectric measurements in the frequency range from 10^{-2} Hz to $1.8 \cdot 10^9$ Hz can be found elsewhere [9,10,19–21].

Isothermal data of the dielectric loss ϵ'' are fitted to a superposition of a relaxation function according to Havriliak-Negami (HN) and a conductivity contribution [32].

$$\epsilon'' = \frac{\sigma_0}{\epsilon_0} \cdot \frac{a}{\omega^s} - Im\left[\frac{\Delta\epsilon}{(1 + (i\omega\tau)^\alpha)^\gamma}\right] \quad (1)$$

In this notation ϵ_0 is the vacuum permittivity, σ_0 the DC-conductivity, $\Delta\epsilon$ the dielectric strength. α and γ describe the symmetric and asymmetric broadening of the relaxation peak. The exponent s equals one for Ohmic behaviour, deviations ($s < 1$) are caused by electrode polarization effects, a is a factor having the dimension sec^{1-s}. The accuracy in the determination of $\log_{10}\tau$ is ≤ 0.1 decades and of $\Delta\epsilon \leq 5\%$. Due to the fact that ϵ' and ϵ'' are connected by the Kramers-Kronig relations a fit in ϵ' would not improve the accuracy of the data analysis. From the fits according to equation (1) the relaxation rate $1/\tau_{max}$ can be deduced which is given at the frequency of maximum dielectric loss ϵ'' for a certain temperature. A second way to interpret the data is the use of a relaxation time distribution of an ensemble of Debye relaxators with relaxation times τ and a distribution density $g(\tau)$. The imaginary part of the dielectric function is then given by

$$\epsilon'' = (\epsilon_{st} - \epsilon_\infty) \int \frac{g(\tau)}{1 + \omega^2\tau^2} d\tau \qquad (2)$$

where ϵ_{st} and ϵ_∞ denote the low and high frequency of the permittivity. $g(\tau)$ can be calculated analytically from the data [33] or extracted from the fit with HN-functions [34]. To characterize the temperature dependence of the relaxation behaviour an averaged relaxation time τ_{med} is used:

$$\log \tau_{med} = \langle \log \tau \rangle = \frac{\int_{-\infty}^{+\infty} \log \tau \cdot g(\log \tau) d\log \tau}{\int_{-\infty}^{+\infty} g(\log \tau) d\log \tau} \qquad (3)$$

τ_{med} equals τ_{max} if the peak of a relaxation process is broadened only symmetrically. In general τ_{med} contains information about the low and high frequency side of a relaxation process whereas τ_{max} denotes only the position of the maximum loss. On the other hand, the calculation of τ_{med} can be done only with high accuracy if the dielectric strength is comparably strong. For that reason τ_{max} is shown for molecules confined to zeolites or MCM-materials and τ_{med} is calculated if the nanoporous sol-gel glasses are used.

Results and Discussion

Propylene Glycol in MCM-41 and MCM-48

The dielectric measurements are carried out with MCM-41 and MCM-48 both having a pore diameter of 2.7 nm. The spectra of propylene glycol in MCM-materials show two relaxation processes which are well separated in frequency. As the relaxation rates of the fast process are comparable to the bulk liquid at high temperatures, this process is assigned to the dynamic glass transition (α-relaxation) in the mesoporous environment. An additional loss process is caused by a Maxwell-Wagner-polarization [35,19,18,21].

Propylene glycol shows (Fig. 2) in the uncoated pores of MCM-41 and MCM-48 a surface effect. The molecular dynamics is shifted to lower

272

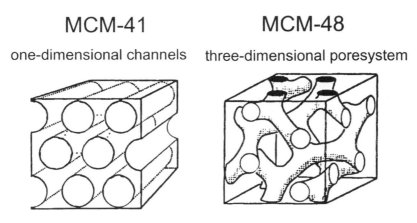

Figure 1. Scheme of the pore system in MCM-41 (left) and MCM-48 (right).

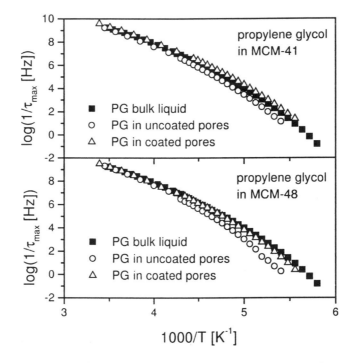

Figure 2. Mean relaxation rate versus inverse temperature of propylene glycol as bulk liquid (solid squares) and confined to uncoated (open circles) and silanized pores (triangles) of MCM-41 (top) and MCM-48 (bottom) having a pore diameter of 2.7 nm. The errors are smaller than the size of the symbols. The Maxwell-Wagner-polarization is omitted.

values compared to the bulk liquid due to the formation of hydrogen-bonds between the propylene glycol molecules and the hydrophilic silica pore walls of the M41S materials [18]. After hydrophobization of the silica walls the formation of hydrogen bonds to the pore surface is hindered (see Figure 3).

The suppression of the surface effect results in a relaxation rate of propylene glycol in the coated mesoporous hosts which is comparable to the bulk liquid in the whole temperature range for MCM-41 and MCM-48 samples [21]. From the size of the pores one has to conclude that the molecular rearrangements of PG take place on a length scale of $\leq 2\,$nm.

Ethylene Glycol in Zeolites

Figure 4 shows the dielectric spectra for ethylene glycol (EG) confined to different zeolitic host systems at 160 K. The relaxation rates τ_{max} for EG in the zeolitic host systems differ by many orders of magnitude: In zeolites with smaller pores (silicalite and sodalite) the relaxation rates of EG are significantly faster compared to zeolite beta and $AlPO_4$-5. Especially for EG in sodalite, the relaxation strength is comparably weak. This is caused by EG molecules which are immobilized due to the interaction with the zeolitic host matrix.

Figure 5 shows the logarithmic relaxation rate as a function of the inverse temperature for EG as bulk liquid and confined to different zeolites. EG in zeolite beta (open triangles) and in $AlPO_4$-5 (open inverted triangles) has a relaxation rate like the bulk liquid (squares) showing a Vogel-Fulcher-Tammann- (VFT-) temperature dependence [36–38] which is characteristic for glass forming liquids:

$$\frac{1}{\tau} = A \exp\left(\frac{-DT_0}{T - T_0}\right) \tag{4}$$

In this notation A is a prefactor, D the fragility parameter and T_0 is the Vogel temperature.

The relaxation rates of EG in silicalite (solid circles), H-ZSM-5 (open circles) and sodalite (stars) show an Arrhenius-type temperature dependence. The single-molecule relaxation of EG in sodalite is at $T \approx 155\,$K by about six orders of magnitude faster compared to the bulk liquid. Its activation energy is $26\pm1\,$kJ/mol and corresponds to the value for bulk EG at high frequencies and high temperatures ($29\pm2\,$kJ/mol) [39]. Extrapolating the relaxation rate of the single molecules in sodalite to higher temperatures leads to the values of the bulk liquid. Hence one has to conclude that for high temperature the relaxation of the EG molecules even in the bulk

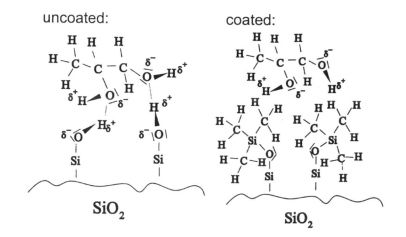

uncoated: coated:

SiO$_2$ SiO$_2$

Figure 3. Schematic diagram of propylene glycol in the neighborhood of an uncoated SiO$_2$ surface (left) and a silanized SiO$_2$ surface (right).

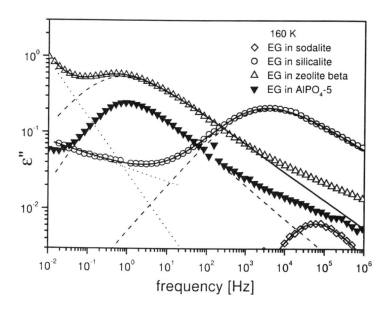

Figure 4. Dielectric loss ϵ'' versus frequency for ethylene glycol (EG) being confined to zeolitic host systems as indicated. The solid lines are fits to the data (dotted line: HN-relaxation, dashed line: conductivity contribution). A deviation in the high frequency wing as for EG in zeolite beta has been observed for bulk liquids as well [41].

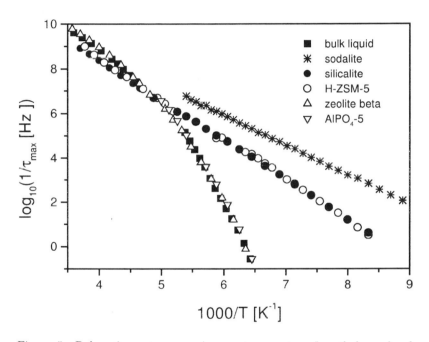

Figure 5. Relaxation rate versus inverse temperature for ethylene glycol being confined to different zeolitic host systems as indicated. The errors are smaller than the size of the symbols.

Table 1. Distance between molecules, average length of hydrogen-bonds (O-H$\cdots$O bonds with a length up to 0.3 nm) and density as calculated from the molecular simulations for ethylene glycol confined to zeolite beta and silicalite and for the bulk liquid. For simulation of the bulk liquid a limited volume (6.64 nm^3) was filled with EG molecules until the bulk density of 1.113 g/cm^3 was reached. In contrast the densities of EG confined in zeolites are results of the simulation.

	Distance between molecules [nm]	Average length of H-bonds [nm]	Density [g/cm^3]
Bulk liquid	0.42±0.01	0.23±0.02	1.113
Silicalite	0.42±0.01	0.24±0.02	1.0±0.1
Zeolite beta	0.41±0.01	0.25±0.02	1.0±0.1
AlPO$_4$-5	0.42±0.01	0.24±0.02	1.2±0.1

has an intramolecular origin. The relaxation rates of EG in silicalite and in H-ZSM-5 are identical and have a larger activation energy (35±2 kJ/mol) which is still smaller than the apparent activation energy (tangent to the VFT-temperature dependence) of the bulk liquid. Its Arrhenius-like temperature dependence resembles the single-molecule relaxation of EG in sodalite and H-ZSM-5.

To complete the information about the guest molecules in the zeolitic host systems computer simulations were carried out to study the molecular arrangement of the molecules in the confining space. Details are described in earlier publications [9,10]. For EG in zeolitic host systems one finds that in silicalite the molecules are aligned almost single-file-like along the channels and that in zeolite beta and in AlPO$_4$-5 two EG molecules are located side by side in the channels (Figure 6). In sodalite the molecules are separated from each other.

Further analysis of the computer simulation shows that neither for the distance between molecules nor for the average length of hydrogen-bonds or for the density a significant change is found between the bulk liquid and the molecules in the restricting geometry (Table 1). But the different host/guest-systems can be distinguished by the average number of neighboring guest molecules (coordination number) (Figure 7).

The coordination number of 11 corresponds to the maximum value in the case of the random close packing model [40] and is found for the bulk liquid within a radius of r=0.66 nm. EG in zeolite beta and in AlPO$_4$-5 has

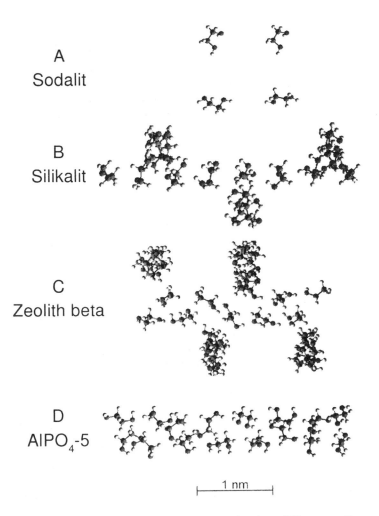

A
Sodalit

B
Silikalit

C
Zeolith beta

D
AlPO$_4$-5

├─────── 1 nm ───────┤

Figure 6. Arrangement of EG molecules confined to different zeolites as obtained by computer simulations. For better visibility the framework of the zeolites is omitted. Picture A shows the distance between the EG molecules in sodalite. The pictures B,C,D show the arrangement of EG molecules in one channel of silicalite, zeolite beta and AlPO$_4$-5, respectively.

only five neighboring molecules showing the same molecular dynamics. In AlPO$_4$-5 only two molecules are located side by side in the one dimensional channels, hence the interactions are dominated by the nearest neighboring molecules and an ensemble as small as six EG molecules is sufficient to show a liquid-like dynamics. AlPO$_4$-5 has no intersections between its channels in contrast to zeolite beta whose intersections contribute much to the pore volume. For this reason the topology of the host systems plays a minor role for the dynamics of H-bonded guest molecules. Further reduction in the channel size (as in the case of silicalite or H-ZSM-5) decreases the average number of neighboring molecules by about 1. This results in a sharp transition from a liquid-like dynamics to that of single molecules.

Salol in Lubricated Nanoporous Sol-Gel Glasses

The molecular dynamics of the glass transition of the "quasi"- van der Waals liquid salol confined to nanopores (2.5, 5.0 and 7.5 nm) with lubricated inner surfaces is found to be faster (by up to two orders of magnitude) than in the bulk liquid (Figure 8 [20]).

This confinement effect is also supported in calorimetric measurements where the most pronounced decrease of the glass transition temperature is found for the liquid contained in the smallest pores [20]. The question arises if this finding could be explained by a (pore-size dependent) decrease of the density of the confined glassy system. Assuming this conjecture for the relaxation rate of the host system a VFT-dependence has to be expected, where the Vogel temperature varies with the pore size in a similar way as the measured glass transition temperature (Figure 9a).

Then the derivative $d(\log(1/\tau))$ / $d(1000/T)$ delivers a temperature dependence as shown in Figure 9b. This can be compared with the *difference* quotient as determined *experimentally* from the relaxation rate measured in temperature steps of 0.5 K (Figure 9c). It turns out that the experimentally deduced difference quotient behaves qualitatively different in its temperature dependence compared to the calculated differential quotient in Figure 9b. In the temperature interval between 333 K and 260 K the apparent activation energies for the bulk and the confined (2.5 and 7.5 nm) liquid coincide within experimental accuracy. But for lower temperatures suddenly the charts bend off; this takes place for the 2.5 nm pores at 256±3 K and for 7.5 nm pores at 245 K±3 K . The temperature dependence is in sharp contrast to the results (Figure 9b) which one would have to expect from a dependence as displayed in Figure 9a, assuming a slowly varying temperature dependence of the density. Hence one has to conclude that the experimental observations cannot be explained on the basis of such a

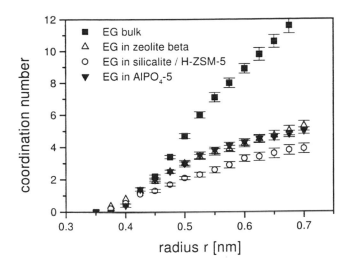

Figure 7. Average number of neighboring molecules (coordination number) as a function of the radius of a surrounding sphere as calculated from the simulations for EG bulk liquid (squares), EG confined to silicalite/H-ZSM-5 (circles), to zeolite beta (open triangles) and to AlPO$_4$-5 (solid triangles).

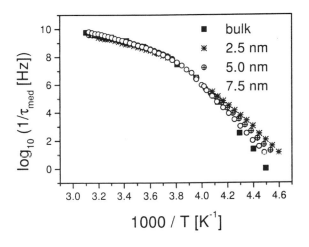

Figure 8. Mean relaxation rate versus inverse temperature of salol as bulk liquid (solid squares) and confined to silanized pores of sol-gel glass having different mean diameters.

(Reproduced with permission from reference 10. Copyright 1999 IDP.)

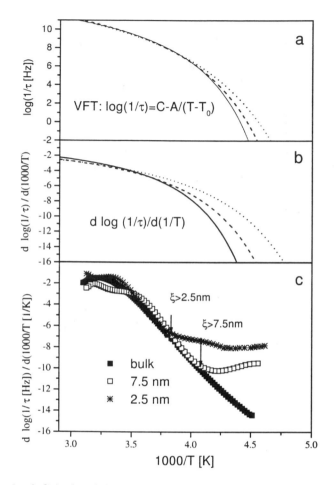

Figure 9. a) Calculated dependence of the mean relaxation rate versus inverse temperature assuming a VFT-law, where T_0 is shifted according to the calorimetrically measured shift of T_g. Solid line: bulk liquid $T_g = 222$ K; dashed line: salol confined to a silanized sol-gel glass having a mean diameter of 7.5 nm: $T_g = 214$ K; dotted line: salol confined to a silanized sol-gel glass having a mean diameter of 2.5 nm: $T_g = 207$ K. b) Calculated derivative of $log(1/\tau[Hz])$ versus inverse temperature assuming the temperature dependence shown in a). The units are omitted due to graphical reasons. They are the same like in part c) of this Figure. c) Experimentally determined difference quotient $\Delta(log(1/\tau[Hz]))/\Delta(1000/T[K])$ for the data shown in Figure8 for the bulk liquid (solid squares), salol confined to silanized nanopores of 7.5 nm (open squares) and salol confined to silanized nanopores of 2.5 nm (stars).

(Reproduced with permission from reference 10. Copyright 1999 IDP.)

density argument. Instead it is suggested that the measured confinement effects are caused by the cooperative nature of the dynamic glass transition. With decreasing temperature the size of cooperative rearranging domains is growing and the apparent activation energy increases. If due to the confinement of the nanoporous system a further growth is prohibited the VFT-dependence turns suddenly into an Arrhenius-like type of thermal activation.

Conclusion

Broadband dielectric spectroscopy (10^{-2}-10^9 Hz) is employed to study the molecular dynamics of low molecular glass forming liquids (propylene glycol, ethylene glycol, salol) in different confining host systems (MCM-materials, zeolites, nanoporous sol-gel glasses). The results are interpreted on the basis of a counterbalance between surface- and confinement effects. While the former decreases the molecular dynamics (causing an increase of the calorimetric glass transition temperature) the latter causes a strong increase of the relaxation rate of the confined liquid (and hence a pronounced decrease of the (calorimetric) glass transition temperature). The observed findings can be explained only on the basis of the cooperative nature of the dynamic glass transition.

Acknowledgements

Support by the Deutsche Forschungsgemeinschaft within the "SFB 294" and the Schwerpunktprogramm "Nanostrukturierte Wirt/Gast-Systeme" is gratefully acknowledged. The authors thank P. Behrens, W. Schwieger, Ö. Weiß, F. Schüth and J. Kärger for providing the samples.

References

1. Angell, C. A. *Science* **1995**, *267*, 1924.
2. Stillinger, F. H. *Science* **1995**, *267*, 1935.
3. Frick, B.; Richter, D. *Science* **1995**, *267*, 1939.
4. Böhmer, R.; Ngai, K. L.; Angell, C. A.; Plazek, D. J. *J. Chem. Phys.* **1993**, *99*, 4201.
5. Hansen, J. P. *Theory of Simple Liquids;* Academic Press: London, 2 ed.; 1990.

6. Götze, W. *J. Phys.: Condens. Matter* **1999**, *11*, A1-A45.

7. Cummins, H. Z. *J. Phys.: Condens. Matter* **1999**, *11*, A95-A117.

8. Ngai, K. L. *J. Phys.: Condens. Matter* **1999**, *11*, A119-A130.

9. Huwe, A.; Kremer. F.; Behrens, P.; Schwieger, W. *Phys. Rev. Lett.* **1999**, *82*, 2338.

10. Kremer. F.; Huwe, A.; Arndt, M.; Behrens, P.; Schwieger, W. *J. Phys.: Condens. Matter* **1999**, *11*, A175-A188.

11. Adam, G.; Gibbs, J. H. *J. Chem. Phys.* **1965**, *43*, 139.

12. Donth, E. *Glasübergang;* Akademie Verlag: Berlin, 1981.

13. Götze, W. . In *Liquids, Freezing and Glass Transition;* Levesque. D.; Hansen, J. P.; Zinn-Justin, J., Eds.; North-Holland: Amsterdam, 1991.

14. Ngai, K. L. Universal Patterns of Relaxation in Complex Correlated Systems. In *Disorder Effects on Relaxational Processes;* Richert, R.; Blumen, A., Eds.; Springer-Verlag: Berlin, 1994.

15. Sappelt, D.; Jäckle, J. *J. Phys. A* **1993**, *26*, 7325.

16. Fischer, E. W.; Donth, E.; Steffen, W. *Phys. Rev. Lett.* **1992**, *68*, 2344.

17. Fischer, E. W. *Physica A* **1993**, *201*, 183.

18. Gorbatschow, W.; Arndt, M.; Stannarius, R.; Kremer, F. *Europhys. Lett.* **1996**, *35*, 719.

19. Arndt, M.; Stannarius, R.; Gorbatschow, W.; Kremer, F. *Phys. Rev. E* **1996**, *54*, 5377.

20. Arndt, M.; Stannarius, R.; Groothues, H.; Hempel, E.; Kremer, F. *Phys. Rev. Lett.* **1997**, *79*, 2077.

21. Huwe, A.; Arndt, M.; Kremer, F.; Haggemüller, C.; Behrens, P. *J. Chem. Phys.* **1997**, *107*, 9699.

22. Huwe, A.; Kremer, F.; Hartmann, L.; Kratzmüller, T.; Braun, H.; Kärger. J.; Behrens, P.; Schwieger, W.; Ihlein, G.; Weiß, O.; Schüth. F. *J. Phys. IV* **2000**, *10*, Pr7-59.

23. Bibby, D. M.; Dale, M. P. *Nature* **1985**, *317*, 157.

24. Donth, E. *Relaxation and Thermodynamics in Polymers, Glass Transition;* Akademie Verlag: Berlin, 1992.

25. Kresge, C. T.; Leonowicz, M. E.; Roth, W. J.; Vartuli, J. C.; Beck, J. S. *Nature* **1992**, *359*, 710.

26. Beck, J. S.; Vartuli, J. C.; Roth, W. J.; Leonowicz, M. E.; Kresge, C. T.; Schmitt, K. D.; Chu, C. T. W.; Olson, D. H.; Sheppard, E. W.; McCullen, S. B.; Higgins, J. B.; Schlenker, J. L. *J. Am. Chem. Soc.* **19**, *114*, 10834.

27. Behrens, P.; Stucky, G. D. *Angew. Chem.* **1993**, *105*, 729.

28. Behrens, P.; Glaue, A.; Haggenmüller, C.; Schechner, G. *Solid State Ionics* **1997**, *101-103*, 255.
29. Braunbarth, C. M.; Behrens, P.; Felsche, J.; van de Goor, G. *Solid State Ionics* **1997**, *101-103*, 1273.
30. Newsam, J. M.; Treacy, M. M. J.; Koetsier, W. T.; de Gruyter, C. *Proc. Roy. Soc. (London)* **1988**, *420*, 375.
31. Meier, W. M.; Olson, D. H.; Baerlocher, C. *Atlas of Zeolite Structure Types;* Elsevier: Amsterdam, 1996.
32. Havriliak, S.; Negami, S. *J. Polym. Sci. Part C* **1966**, *14*, 99.
33. Schäfer, H.; Sternin, E.; Stannarius, R.; Arndt, M.; Kremer, F. *Phys. Rev. Lett.* **1996**, *76*, 2177.
34. Havriliak, S.; Negami, S. *Polymer* **1967**, *8*, 161.
35. Schüller, J.; Mel'nichenko, Y. B.; Richert, R.; Fischer, E. W. *Phys. Rev. Lett.* **1994**, *73*, 2224.
36. Vogel, H. *Phys. Zeit.* **1921**, *22*, 645.
37. Fulcher, G. S. *J. Am. Chem. Soc.* **1925**, *8*, 339.
38. Tammann, G.; Hesse, G. *Anorg. Allgem. Chem.* **1926**, *156*, 245.
39. Jordan, B. P.; Sheppard, R. J.; Szwarnowski, S. *J. Phys. D* **1978**, *11*, 695.
40. Cusack, N. E. *The Physics of structurally disordered Matter;* Adam Hilger: Bristol, 1987.
41. Hofmann, A.; Kremer, F.; Fischer, E. W.; Schönhals, A. The Scaling of the α- and β-Relaxation in Low Molecular Weight and Polymeric Glassforming Systems. In *Disorder Effects on Relaxational Processes*; Richert, R.; Blumen, A., Eds.; Springer-Verlag: Berlin, 1994.

Chapter 21

Size-Dependent Dielectric Properties of Liquid Water Clusters

Joel E. Boyd[1], Ari Briskman[1], Alan Mikhail[1], Vicki Colvin[1], and Daniel Mittleman[2,*]

Departments of [1]Chemistry and [2]Electrical and Computer Engineering, Rice University, 6100 South Main Street, Houston, TX 77251

A large absorption peak has been observed in the far infrared spectrum of water confined within inverse micelles of sodium bis(2-ethylhexyl) sulfosuccinate (AOT) in heptane. The amplitude and spectral position of this peak depend on the size of the water pool. The proposed origin of these spectral features lies in surface oscillations of the water pool. The presence of a large excess in the vibrational density of states of confined water could have far-ranging implications in many biochemical and chemical processes where confined water is present.

Introduction

The properties of water confined on the nanometer scale are important in a diverse range of disciplines. Many processes in biology and chemistry occur within water cavities of nanometer-sized dimensions. Biological macromolecules, most notably proteins, not only function in aqueous environments, but also contain large fractions of water within their structure. This "bound" water can consist of only a few water molecules or more than several hundred. Its presence is crucial for many functions of biological systems (54,27,5,60,57,61,49,20,9,7), yet the most basic properties, such as density and dielectric constant, cannot readily be characterized *in-situ*. Confined water is also frequently encountered in areas of materials processing. Many high surface area catalysts possess small cavities where chemical reactions are accelerated (2,10,29); quantitative predictions for reaction rates in these systems require some estimate of the solvent's properties. Nanoscale reactors defined by inverse micelles have become one popular route to the formation of nanocrystalline materials (48,50,40,24). As it becomes more important to effect quantitative control over reaction rates in these media, many fundamental properties of the water in these environments will need to be characterized.

One such property that is of interest to many scientific fields is the frequency-dependent dielectric function of confined water. The dielectric function of bulk water exhibits its primary relaxation features in the far-infrared region, between 1 and 100 cm^{-1}. This is therefore an important spectral region for dielectric characterization of confined water. Terahertz time-domain spectroscopy (TDS) allows for the measurement of both the real and imaginary parts of the dielectric function over much of this range. Such measurements provide a direct indication of the vibrational density of states of the liquid. Other time-domain spectroscopies which measure the low frequency solvent states do so indirectly through the behavior of probe molecules. While these methods can identify those modes active in solvation, they do not provide the full spectrum of solvent states. Terahertz spectroscopy fills an important niche between these other experimental approaches to the study of liquid state dynamics. Here we report the far-infrared dielectric behavior of confined water along with discussion of its relevance and implications.

One field in which the far-infrared study of confined water can have a direct impact is the dynamics of glass-forming liquids. Upon cooling, many liquids develop an anomalous peak, known as the Bose peak, in their vibrational density of states in the far-infrared. The microscopic origins of this feature are the subject of much debate. The presence of a Bose peak has been correlated with numerous other properties of disordered materials, such as the well-known low-temperature specific heat anomaly in glasses (18,39). Further, the amplitude of the Bose peak appears to be strongly correlated with the glass former's fragility, a useful phenomenological description based on macroscopic properties such as viscosity (4,15,22). While bulk water does not exhibit a Bose peak, the confined water in the experiments reported here, as well as very recent neutron scattering studies of the bound water associated with proteins (45), shows evidence for such a feature.

The inverse micelles used in this work provide three-diminsional confinement of water with well-controlled size distributions and evident biological relevance. Inverse micelles with AOT, the sodium salt of Bis(2-ethylhexyl)sulfosuccinate, can be formed with water pools ranging from essentially zero size to as many as 100,000 water molecules, and have been extensively characterized so that size dispersion, interfacial organization, and interior water structure are well established (32,63,52,59,62,36). One drawback to using inverse micelles for investigations of far infrared dynamics is that they do not lend themselves well to temperature-dependent studies, because the micelle size and stability are dependent upon temperature (63). The wide range and controllability of their sizes, however, make them extremely attractive for studying the length scales of collective vibrational modes of water.

Experimental

Materials

AOT (99%) was purchased from Sigma and purified based on the method of Martin and Magid (37). The purified AOT was then kept over P_2O_5 (Aldrich) in a vacuum dessicator at least 48 hours before use. Karl Fischer titrations revealed that the residual water content after this process was less than 0.2 moles of water per mole of AOT.

Anhydrous heptane (Aldrich, 99%) was used as received. Samples were prepared by dissolution of known weights of AOT in 30 ml of heptane. Millipore water (18.2 MΩ/cm^2) was then injected to make samples of varying w, the molar ratio of water to surfactant in solution (w=[water]/[AOT]), and volume fraction of 6%. Samples were used after stirring overnight and within 4 days of preparation to minimize hydrolysis of the surfactant (21,17,31).

Methods

FTIR measurements were performed on a Nicolet Magna IR 760 spectrometer. The OH stretch of water centered at ~3500 cm^{-1} is broad, and can be analyzed according to standard methods established in the literature (12,42). The peak is fit to the sum of 3 gaussian peaks assigned to be "free" (3300cm^{-1}), "bound" (3450 cm^{-1}), and "trapped" (3600 cm^{-1}) water in the micellar solution.

The far-infrared dielectric spectroscopic data were collected with a relatively new method, terahertz time-domain spectroscopy (TDS) (30,53,38). This method provides both the absorption and refractive index of a sample without resorting to a Kramers-Kronig analysis. The far-IR (THz) radiation is coherently emitted and detected using femtosecond (fs) pulses from a 80 MHz titanium sapphire laser (Coherent® Mira) operating at 789 nm. These ultrashort pulses are focused onto antennae that consist of a pair of gold leads photolithographically deposited onto semiconductor (GaAs) substrates. The transmitter antenna is biased at 20 volts and the acceleration of the electrons excited above the bandgap by the fs pulse results in the emission of broadband radiation from ~0.1 to 1.5 THz. The THz pulse is then collimated by a combination of a hyperhemispherical silicon lens mounted directly onto the substrate, and a high-density polyethylene lens. After passing through the sample, the THz beam is refocused onto the receiving lens. The charge carriers created by the fs optical excitation within the receiver antenna gap are accelerated by the incoming THz field. The current measured across the receiving antenna is thus proportional to the amplitude of the THz electric field. The relative delay between the THz pulse and the optical pulse on the receiver can be varied to measure the waveform as a function of time via photoconductive sampling. The current detected at the receiver is amplified by a current to voltage amplifier and then recorded using a lockin amplifier which is queried by a computer via GPIB interface. A diagram of the spectrometer is provided in Figure 1.

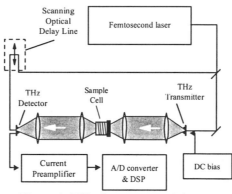

Figure 1: THz Spectrometer Diagram

The sample is contained in a polyethylene bag which is inserted into a variable pathlength cell comprised of two high-density polyethylene windows;

one of these windows is fixed, and the other is attached to a computer-controlled stepper motor. A complete scan is collected at each of a series of 7 pathlengths varying from 5mm to 1mm in thickness. The multiple pathlength data is then used to calculate both the refractive index (n) and absorption (α) of the sample, using the procedure outlined in reference (30). The spectral range over which these quantities can be determined depends on the bandwidth of the radiation source and the absorption of the sample. Data points at frequencies where the amplitude of the signal is not 20 times greater than that of the noise spectrum are excluded.

The data measured from a sample contains the aggregate dielectric of the entire sample, including contributions from the heptane, the surfactant, and the water. In order to access the dielectric response of the water alone, a series of deconvolutions are performed according to the core-shell dielectric model developed by Hanai *et al.* (23). If the dielectric of the background solvent is known, then the dielectric of the micelle is determined from:

$$\frac{\varepsilon_{solution} - \varepsilon_{micelle}}{\varepsilon_{background} - \varepsilon_{micelle}} \cdot \left(\frac{\varepsilon_{background}}{\varepsilon_{solution}}\right)^{1/3} = 1 - \varphi$$

where φ is the volume fraction of micelles in solution. From $\varepsilon_{micelle}$, one can determine the dielectric of the core if the dielectric of the surfactant shell is known, using:

$$\varepsilon_{micelle} = \varepsilon_{shell} \cdot \frac{2(1-v)\varepsilon_{shell} + (1+2v)\varepsilon_{core}}{(2+v)\varepsilon_{shell} + (1-v)\varepsilon_{core}}$$

The dielectric of the shell is determined experimentally, by measuring the dielectric of a solution containing AOT, but no water. In this case, it is well established that the surfactant still forms spherical aggregates, with the polar head groups at the center (63). These w=0 samples are formulated to have the same volume fraction of AOT as the corresponding hydrated micellar sample (Figure 2A). Here, v is the volume fraction of the water core within the micelle, which is determined using literature values for the thickness of the AOT shell (36) and the water pool radius (63). The result is ε_{core}, the dielectric of the water in the interior of the micelle, from which the absorption and refractive index may be derived.

Results and Discussion

FTIR

The water contained within inverse micelles is generally regarded to be of two distinct types: interfacial water which is strongly perturbed by interactions with the surfactant head groups, and interior water which has a hydrogen-bonded structure nearly identical to that of bulk water. The NMR proton chemical shift provides information about the local structure of the hydrogen bond network. Such studies indicate that interfacial water is perturbed by the surfactant interface (62,34,36,26,33). Infrared spectroscopy of the OH-stretch region can also provide insight into the hydrogen bond network of water within inverse micelles. Higher-lying components (at ~3450 cm^{-1} and ~3600 cm^{-1}) have been assigned to molecules in more distorted configurations and to non-H-bonded molecules, respectively (19,42). In very small micelles, these higher frequency components dominate the spectrum. (12,42,8,35,11,16) IR studies have shown that the interfacial water layer is about two monolayers thick, in agreement with molecular dynamics simulations (14). Thus, in very small micelles, all of the internal water can be classified as "interfacial" in nature. Once inverse micelles exceed ~15 Å in radius, the water in the central core begins to develop a local hydrogen bond network spectroscopically identical to bulk water. In both the bulk liquid and in reverse micelles, one observes an OH stretch three-component lineshape, of which the lowest-lying component (at ~3330 cm^{-1}) arises from molecules in regular, unstrained H-bond configurations. As the AOT micelle is made larger, the relative contribution of the free water peak grows (Figure 2) (12). Onori and Santucci have proposed an equilibrium balance between the free water pool and the bound water layer residing within a few angstroms of the polar headgroups of the surfactant (42). Proton NMR spectroscopy supports the model of an interior bulk-like water pool whose properties approach those of the bulk liquid at water pool radii greater than 15 Å (62,34,36,26,33).

Another complicating factor with the use of AOT, is the presence of a sodium counterion in the interior of the inverse micelle. Na-NMR studies found that the Na$^+$ mobility is dramatically reduced relative to bulk solutions. This has been interpreted as indicating that the ion is confined in the bound water layer, close to the AOT headgroups, and not solvated in the free water core (62,34). These results are consistent with recent molecular dynamics simulations (14) and the conclusions of fluorescence probe measurements (51). Further experimental results regarding this issue are presented below.

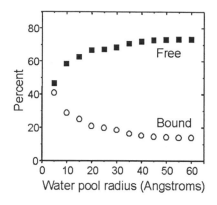

Figure 2: Relative fractions of free and bound water in AOT inverse micelles, determined via line shape analysis of the FTIR spectra of the OH-stretch region (12).

Terahertz Spectroscopy

In addition to the structural studies of the hydrogen bonded network in inverse micelles, terahertz TDS provides information on the dynamics of the confined water. In contrast to the relatively featureless absorption profile of bulk water in the far-infrared, the absorption spectrum of the confined water in this study exhibits a pronounced peak when the micellar radius ranges from 15 Å to 45 Å. Figure 3 shows the absorption spectra of several representative samples, demonstrating the presence of a marked peak in the smaller water pools. The larger micelles studied do not exhibit the absorption peak, and resemble the absorption spectrum of bulk water. The similarities between the flat spectrum of the large micelles and the spectrum of bulk water (figure 3) seem to indicate the presence of a water pool in the largest micelles that is bulk-like in its dynamical behavior. Furthermore, this bulk-like limit provides support for the validity of the dielectric deconvolution models described earlier. It should be pointed out that the actual magnitude of the absorption spectrum of the largest micelles is not identical to that of the bulk liquid. This could reflect a change in the limiting behavior of the micellar water similar to that observed in the compressibility and viscosity experiments on AOT inverse micelles by a number of researchers (6,1,3,25). Whatever the explanation for this shift, it remains significant that the spectrum of the largest micelles does not contain an absorption peak, and indeed is featureless and reminiscent of bulk water. Figure 3B is provided to verify that the feature is indeed a single resonance. For a resonant process, this Cole-Cole plot should be circular (55); the lines show best fits to circles, demonstrating excellent agreement. The presence of such a large terahertz resonance has no precedent in bulk water, or even most bulk liquids at normal temperatures. Its presence in inverse micelles, and its strong size dependence suggest that it arises from the restricted dimensions of the micelle cavity.

The spectral position and amplitude of the absorption peak are dependent upon the size of the water pool inside the micelle. As the micelles become larger, the peak decreases in amplitude and shifts to lower frequencies. Figure 4 shows the position of the peak as a function of water pool size. The striking nature of this size dependence indicates that the peak is probably not due merely to interfacial effects since the fraction of interfacial (bound) water does not increase substantially in this size range (36,42) (Figure 2).

One possible explanation for the appearance of this peak is the presence of the Na^+ counterion within the water pool. To investigate this possibility, inverse micelles were formulated with additional Na^+ ions added in the form of NaCl in concentrations varying from .1 to 1M Na^+ in the water pool. In all of these salt-added samples, the absorption peak was still present. Samples with the highest salt concentrations possessed peaks that were shifted to lower frequencies. It is quite likely that this peak shift is due to the expected change in size of an AOT micelle where brine is used in place of water (56). The presence of additional Na^+ (and Cl^-) ions did not disrupt the presence of the absorption peak, and it is further unlikely that the native Na^+ ions could be the cause of this effect since they have been demonstrated to be electrostatically associated with the surfactant headgroups, as discussed earlier. Further experiments involving other surfactants, both nonionic and involving other counterions are in progress.

The most striking aspect of this far infrared absorption peak, its size-dependent spectral position, can be accounted for within a model for the vibrational mode spectrum of small liquid droplets developed by Tamura *et al.* (58) The eigenmodes in this model divide into two distinct categories, one which represents surface vibrations (shape oscillations) and another which involves internal vibrations (compressional oscillations). These latter modes fall at frequencies larger than $\pi v_c/R$, where v_c is the speed of sound in the liquid and R is the radius of the spherical droplet. Using the bulk value for the sound velocity, one finds that the majority of these internal modes lie at frequencies above 1 THz, too high to explain the resonances observed. In contrast, the surface modes are lower frequency excitations, closer to the range of the observed modes. Surface modes are harmonic shape oscillations arising from the restoring force associated with the interfacial tension at the micelle surface. The only parameters involved in calculating the eigenfrequencies of these surface modes are the water pool radius, the densities of water and heptane, and the interfacial tension of the water pool. The size-dependent interfacial tension ($\sigma(R)$) of an AOT micelle has been considered in great detail by Peck and co-

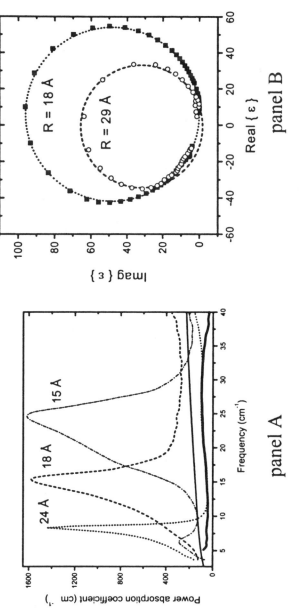

Figure3: *Panel A: Absorption spectra of several micelle samples, with different water pool radii shown. The thick, nearly featureless curve represents a radius of 75 Å, one of the largest micelles studied to date. The thin line above is the curve for bulk water. Panel B: Cole-Cole plots for two samples demonstrating resonant absorption behavior.*

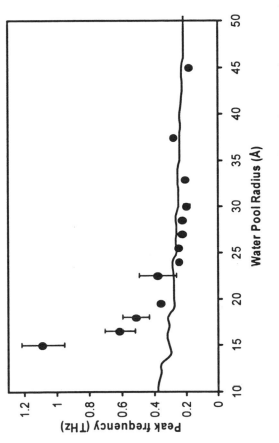

Figure 4: The peak frequency of the absorption peak as a functionof water pool radius. The curve represents the mean value of the calculated density of modes due to surface fluctuations.

workers. (47) Their expression for σ(R) contains contributions from both the water-surfactant and the surfactant-oil interfaces. To extract a numerical value for σ(R), one requires knowledge of several parameters, including the structural parameters for the AOT molecule, the counterion dissociation constant, the effective headgroup and tailgroup areas, and the solubility parameter for heptane. This model predicts a monotonically decreasing σ(R) with increasing radius due mostly to the curvature-dependent electrostatic interaction of the headgroups.

Using this formalism to determine the interfacial tension, one can calculate the vibrational density of mode with no adjustable parameters(58). From these spectra, we may calculate the mean frequency of the surface modes which is compared with the experimental result in figure 4. The correlation of theory with experiment is excellent at large micelle sizes. At smaller radii, the agreement shows a marked deviation from the experimental result. This could be due to a coupling of the surface oscillations to the dielectric relaxational mode of bulk water at 0.9 THz. This coupling would enhance the absorption strength of the surface oscillations and tend to shift them to higher frequencies. This coupling of localized and delocalized vibrational modes would also account for the asymmetric peak shapes shown in figure 5.

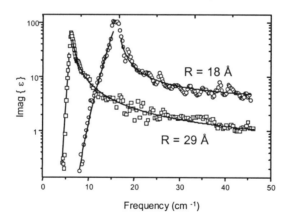

Figure 5: The imaginary part of the dielectric function for two different micelle samples. Solid lines show best fits to power law behavior on the low-frequency edge, and Debye-like behavior on the high frequency edge.

It is worth noting that the observed absorption peaks show some important similarities with another type of far-infrared excitation, namely the Bose peak. This peak is observed in many glasses and glass-forming liquids, and reflects a redistribution of the vibrational density of states resulting from the disorder in the material. The Bose peak has not been observed in supercooled bulk water

(53), but there is evidence that it may exist in confined water. Very recently, several molecular dynamics simulations of confined water have shown evidence of a Bose peak in the calculated dynamical structure factor (28,43,44). Experimentally, the observation of a Bose peak in water has only very recently been reported, and only in one specific system, the hydration shell of a protein (45). The difficulty in making unambiguous observations of a Bose peak in liquid water illustrates the challenge in studying cooperative effects in water without varying the temperature. Thus, the ability to extend this analysis to a size-dependent phenomenon could yield valuable information about the behavior of the water and also about the origins of the Bose peak.

The models developed for the explanation and investigation of the Bose peak in strong glass formers appear relevant to the observed behaviors of the anamolous absorption peak in confined water. One model for the Bose peak is based on the localization of of propagating accoustic modes as a result of strong scattering. The localization occurs when the Ioffe-Regel criterion, that the mean free path of the wave is comparable to its wavelength (13,46,41), is satisfied. Although it is at first difficult to see how this model involving propagating modes and scattering events could be extended to an inherently localised phenomenon like micellar confinement, the Ioffe-Regel criterion does predict essentially the same size dependence exhibited by the absorption peak in figure 4.

The model presented by Novikov (41) predicts a specific form for $\chi''(\omega)$ that correlates well with observed Bose peak spectra; namely that the low frequency edge of the peak should obey power law behavior, and that the high frequency edge should appear Debye-like in character. This model which proposes a coupling between an anharmonic vibration and a Debye-like relaxation, suggests a line shape that is quite similar to the observed anharmonic peak shapes of the confinement-induced peak in micellar water as demonstrated in Figure 5. The value of the exponent of the power law dependence in the confined water resonance (~8) is quite different than that for the Bose peak (0.3 −1). The origins of this spectral line shape remain unclear, although the similarities to the shapes predicted by Novikov suggest a similar coupling mechanism.

Conclusions

The observation of a large absorption feature in the far-infrared spectrum of confined water could prove very important in biological or catalytic processes where confined water is present. A pronounced terahertz resonance, such as the one reported here,- has no precedent in bulk water, or even most bulk liquids at normal temperatures. Its presence in inverse micelles, and its strong size dependence suggest that it arises from the restricted dimensions of the micelle cavity. The marked asymmetry of the observed peaks suggests that a dynamical, rather than a structural, description is required. The surface oscillation model developed for small liquid droplets is capable of predicting the spectral position of the experimental absorption peak in all but the smallest inverse micelles studied. The deviation of the smallest micelles from the model could be due to coupling of these delocalized modes to a localized relaxational mode at 0.9 THz. This coupling would not only account for the theoretical deviation at small sizes, but would also explain the asymmetric experimental peak shapes. The Bose peak in supercooled liquids has extensive similarites to the spectral features of the peak observed in confined micellar water. The size-dependence of this confined water study could prove useful in understanding the dynamics that give rise to the Bose peak.

Acknowledgements

Mr. Jared Turner of Dianal America is gratefully acknowledged for performing the Karl Fischer analysis. This work has been funded by the National Science Foundation, the Robert A. Welsh Foundation, and the Petroleum Research Fund of the American Chemical Society.

References and Notes

(1) Acosta, E.; Bisceglia, M.; Fernandez, J. C. *Coll. Surf. A* **2000**, *161*, 417-422.

(2) Ahmed, S. I.; Friberg, S. *J. Am. Chem. Soc.* **1972**, *94*, 5196-5199.

(3) Amararene, A.; Gindre, M.; Huérou, J.-Y. L.; Urbach, W.; Valdez, D.; Waks, M. *Phys. Rev. E* **2000**, *61*, 682-689.

(4) Angell, C. A. *J. Non-cryst. Sol.* **1991**, *131-133*, 13-31.

(5) Atik, S. A.; Thomas, J. K. *J. Am. Chem. Soc.* **1981**, *103*, 4367-4371.

(6) Bedeaux, D.; Linden, E. V. d.; Dijk, M. A. V. *Physica A* **1989**, *157*, 544-547.

(7) Bellissent-Funel, M.-C. *J. Molec. Liq.* **2000**, *84*, 39-52.

(8) Boicelli, C. A.; Giomini, M.; Giuliani, A. M. *Appl. Spec.* **1984**, *38*, 537-539.

(9) Consta, S.; Kapral, R. *J. Chem. Phys.* **1999**, *111*, 10183-10191.

(10) Corbin, D. R.; Herron, N. *J. Mol. Catal.* **1994**, *86*, 343-369.

(11) D'Angelo, M.; Martini, G.; Onori, G.; Ristori, S.; Santucci, A. *J. Phys. Chem.* **1995**, *99*, 1120-1123.

(12) D'Angelo, M.; Onori, G.; Santucci, A. *Il Nuovo Cimento* **1994**, *16*, 1601-1611.

(13) Elliott, S. R. *Europhys. Lett.* **1992**, *19*, 201-206.

(14) Faeder, J.; Ladanyi, B. M. *J. Phys. Chem. B* **2000**, *104*, 1033-1046.

(15) Ferrer, M. L.; Sakai, H.; Kivelson, D.; Alba-Simionesco, C. *J. Phys. Chem. B* **1999**, *103*, 4191-4196.

(16) Fioretto, D.; Freda, M.; Onori, G.; Santucci, A. *J. Phys. Chem. B* **1999**, *103*, 8216-8220.

(17) Fletcher, P. D. I.; Galal, M. F.; Robinson, B. H. *J. Chem. Soc., Faraday Trans I* **1984**, *80*, 3307-3314.

(18) Flubacher, P. *J. Phys. Chem. Solids* **1959**, *12*, 53-65.

(19) *Water: A Comprehensive Treatise*; Franks, F., Ed.; Plenum Press: New York, 1972; Vol. 1.

(20) García-Río, L.; Leis, J. R.; Iglesias, E. *J. Phys. Chem.* **1995**, *99*, 12318-12326.

(21) Geiger, S.; Eicke, H.-F. *J. Colloid Interface Sci.* **1986**, *110*, 181-187.

(22) Green, J. L.; Ito, K.; Xu, K.; Angell, C. A. *J. Phys. Chem.* **1999**, *103*, 3991-3996.

(23) Hanai, T.; Imakita, T.; Koizumi, N. *Coll. Polym. Sci.* **1982**, *260*, 1029-1034.

(24) Haram, S. K.; Mahadeshwar, A. R.; Dixit, S. G. *J. Phys. Chem.* **1996**, *100*, 5868-5873.

(25) Hasegawa, M.; Sugimura, T.; Shindo, Y.; Kitahara, A. *Coll. Surf. A* **1996**, *109*, 305-318.

(26) Heatley, F. *J. Chem. Soc. Faraday Trans.* **1988**, *84*, 343-354.

(27) Hilhorst, R.; Spruijt, R.; Laane, C.; Veeger, C. *Eur. J. Biochem.* **1984**, *144*, 459-466.

(28) Horbach, J.; Kob, W.; Binder, K.; Angell, C. A. *Phys. Rev. E* **1996**, *54*, 5897-5900.

(29) Johnson, J. W.; Brody, J. F.; Soled, S. L.; Gates, W. E.; Robbins, J. L.; Marucchi-Soos, E. *J. Mol. Catal. A* **1996**, *107*, 67-73.

(30) Kindt, J. T.; Schmuttenmaer, C. A. *J. Phys. Chem.* **1996**, *100*, 10373.

(31) Kotlarchyk, M.; Chen, S.-H.; Huang, J. S.; Kim, M. W. *Phys. Rev. A* **1984**, *29*, 2054-2069.

(32) Kotlarchyk, M.; Huang, J. S.; Chen, S.-H. *J. Phys. Chem.* **1985**, *89*, 4382-4386.

(33) Lang, J.; Mascolo, G.; Zana, R.; Luisi, P. L. *J. Phys. Chem.* **1990**, *94*, 3069-3074.

(34) Lindblom, G.; Lindman, B.; Mandell, L. **1970**.

(35) MacDonald, H.; Bedwell, B.; Gulari, E. *Langmuir* **1986**, *2*, 704-708.

(36) Maitra, A. *J. Phys. Chem.* **1984**, *88*, 5122-5125.

(37) Martin, C. A.; Magid, L. J. *J. Phys. Chem.* **1981**, *85*, 3938-3944.

(38) Mittleman, D. M.; Nuss, M. C.; Colvin, V. L. *Chem. Phys. Lett.* **1997**, *275*, 332-338.

(39) Nakayama, T. *Physica B* **1999**, *263-264*, 243-247.

298

(40) Natarajan, U.; Handique, K.; Mehra, A.; Bellare, J. R.; Khilar, K. C. *Langmuir* **1996**, *12*, 2670-2678.

(41) Novikov, V. N. *Phys. Rev. B* **1998**, *58*, 8367-8378.

(42) Onori, G.; Santucci, A. *J. Phys. Chem.* **1993**, *97*, 5430-5434.

(43) Paciaroni, A.; Bizzarri, A. R.; Cannistraro, S. *Phys. Rev. E* **1998**, *57*, 6277-6280.

(44) Paciaroni, A.; Bizzarri, A. R.; Cannistraro, S. *Physica B* **1999**, *269*, 409-415.

(45) Paciaroni, A.; Bizzarri, A. R.; Cannistraro, S. *Phys. Rev. E* **1999**, *60*, 2476-2479.

(46) Parshin, D. A. *Phys. Rev. B* **1994**, *49*, 9400-9418.

(47) Peck, D. G.; Schechter, R. S.; Johnston, K. P. *J. Phys. Chem.* **1991**, *95*, 9541-9549.

(48) Pileni, M. P. *J. Phys. Chem.* **1993**, *97*, 6961-6973.

(49) Politi, M. J.; Chaimovich, H. *J. Phys. Chem.* **1986**, *90*, 282-287.

(50) Qi, L.; Ma, J.; Cheng, H.; Zhao, Z. *Coll. & Surf. A* **1996**, *108*, 117-126.

(51) Riter, R. E.; Willard, D. M.; Levinger, N. E. *J. Phys. Chem. B* **1998**, *102*, 2705-2714.

(52) Robinson, B. H.; Toprakcioglu, C.; Dore, J. C. *J. Chem. Soc., Faraday Trans. I* **1984**, *80*, 13-27.

(53) Rønne, C.; Thrane, L.; Åstrand, P.-O.; Wallqvist, A.; Mikkelsen, K. V.; Keiding, S. R. *J. Chem. Phys.* **1997**, *107*, 5319-5331.

(54) Samama, J.-P.; Lee, K. M.; Biellmann, J.-F. *Eur. J. Biochem.* **1987**, *163*, 609-617.

(55) Scaife, B. K. P. *Principles of Dielectrics*; Oxford University Press: New York, 1998.

(56) Svergun, D. I.; Konarev, P. V.; Volkov, V. V.; Koch, M. H. J.; Sager, W. F. C.; Smeets, J.; Blokhuis, E. M. *J. Chem. Phys.* **2000**, *113*, 1651-1665.

(57) Swallen, S. F.; Weidemaier, K.; Fayer, M. D. *J. Phys. Chem.* **1995**, *99*, 1856-1866.

(58) Tamura, A.; Ichinokawa, T. *Surface Science* **1984**, *136*, 437-448.

(59) Toprakcioglu, C.; Dore, J. C.; Robinson, B. H. *J. Chem. Soc., Faraday Trans. I* **1984**, *80*, 413-422.

(60) Vos, M. H.; Rappaport, F.; Lambry, J.-C.; Breton, J.; Martin, J. L. *Nature* **1993**, *363*, 320-325.

(61) Weidemaier, K.; Fayer, M. D. *J. Phys. Chem.* **1996**, *100*, 3767-3774.

(62) Wong, M.; Thomas, J. K.; Nowak, T. *J. Am. Chem. Soc.* **1977**, *99*, 4370-4376.

(63) Zulauf, M.; Eicke, H.-F. *J. Phys. Chem.* **1979**, *83*, 480-486.

INDEXES

Author Index

Subject Index

A

Adam–Gibbs concept
 cooperatively rearranging region
  (CRR), 244
 transition probability of cooperative
  rearrangements, 251
Aggregate
 equilibrium distribution of, sizes, 241
 See also Clusters
Aluminophosphate
 description, 270
 See also Molecular dynamics in
  confining geometries
Amorphous solid water (ASW)
 calculated diffusion length versus
  temperature, 208–209
 crystallization time vs. temperature,
  208
 description, 199
 See also Self-diffusivity; Water
Amorphous thin films, lifetime of
 metastable, 208–209
Anomalous dynamics
 behavior of particle center-of-mass
  mean-square displacement, 92–93
 experiment and computer
  simulations, 93
 See also Polymer liquids
Aqueous solution. *See* Electron
 photodetachment in solution; Indole
Au(111) crystal
 picosecond time-resolved X-ray
  diffraction, 77
 shift of rocking curve as function of
  delay time, 82*f*
 temperature gradient in crystal
  lattice, 77, 80
 time-resolved rocking curves, 81*f*

See also Time-resolved X-ray
 diffraction

B

Bead-spring model
 correlation length vs. time, 221*f*
 dynamics of glass-forming liquids
  and polymers, 216–217
 See also Spatially heterogeneous
  dynamics (SHD)
Biological systems, confined liquids,
 8–9
Bose peak
 impact of far-infrared study of
  confined water, 285–286
 models for explanation and
  investigation of, 295
 terahertz spectroscopy, 294–295
2-Butanol. *See* 7-Dehydrocholesterol
 (DHC)

C

Cadmium perchlorate
 addition to indole in 1-propanol,
  129–130
 See also Indole
Caging behavior, molecular motion,
 215–216
Carbon dioxide, liquid
 intermolecular spectroscopy, 34–37
 low frequency intermolecular
  spectroscopy, 40
 time dependence of instantaneous
  normal mode (INM) frequencies,
  32*f*

P

Phenyl salicylate. *See* Salol
Photochemistry
  microscopic understanding in
    solution-phase, 5–6
  polyene chromophores, 148
  See also Chlorine dioxide
    photochemistry
Photodetachment
  aqueous halide anions, 109
  term, 108–109
  See also Electron photodetachment
    in solution
Photoionization
  charge recombination in liquids,
    124–125
  conduction-band electron, 123
  conduction-band photoionization
    (CB-PI) mechanism, 123
  electron-transfer photoionization
    (ET-PI) mechanism, 124
  first ionization potential, 123
  indole for ultrafast solute
    photoionization, 125
  neutral aromatic molecules in polar
    solvents, 125
  neutral molecule, 122–123
  photolysis reactions, 5–6
  sub-conduction-band photoionization
    mechanism, 124
  term, 108–109
  thresholds for indole in various
    solvents, 126*t*
  See also Electron photodetachment
    in solution; Indole
Photoisomerization, polyenes, 6
Photolysis reactions, vs.
  photoionization, 5–6
Photon echo spectroscopy. *See* Third-
  order nonlinear spectroscopy of
  coupled vibrations
Polyenes
  photochemical reactions, 148–149
  photoisomerization, 6

See also 7-Dehydrocholesterol
  (DHC)
Poly(ethylene) (PE)
  comparing simulated data of mean-
    square displacement of single-
    chain center-of-mass and
    monomer mean-square
    displacement for unentangled PE
    melt chains, 99
  simulations of unentangled PE melt
    dynamics, 100, 101*f*
  unentangled undercooled polymer
    melt dynamics, 100, 103
  See also Polymer liquids
Polymer liquids
  anomalous dynamics, 92–93
  appearance of subdiffusive regime,
    93
  comparing theory with simulated
    data of $\Delta R^2(t)$ and monomer mean-
    square displacement, 99
  correlated dynamics generalized
    Langevin equation (CDGLE), 95
  equation of motion, 95
  extended Rouse equation in
    agreement with data, 103–104
  fitting procedure in agreement with
    data, 102*f*
  inconsistencies between Rouse
    equation and experimental data,
    94
  infinitesimal translation of center-of-
    mass, 97
  intermolecular force between two
    tagged chains, 96
  mathematical construction of
    projection operator, 95
  mean-square displacement of single-
    chain center-of-mass, $\Delta R^2(t)$, 97,
    98*f*
  mechanism for temperature decrease,
    103
  non-linear Rouse equations, 96
  optimized intermolecular Rouse
    approximation, 96

318